SpringerBriefs in Mathematics

SpringerBriefs in Mathematics showcases expositions in all areas of mathematics and applied mathematics. Manuscripts presenting new results or a single new result in a classical field, new field, or an emerging topic, applications, or bridges between new results and already published works, are encouraged. The series is intended for mathematicians and applied mathematicians.

BCAM SpringerBriefs

Editorial Board

BCAM *SpringerBriefs* aims to publish contributions in the following disciplines: Applied Mathematics, Finance, Statistics and Computer Science. BCAM has appointed an Editorial Board, who evaluate and review proposals.

Typical topics include: a timely report of state-of-the-art analytical techniques, bridge between new research results published in journal articles and a contextual literature review, a snapshot of a hot or emerging topic, a presentation of core concepts that students must understand in order to make independent contributions.

Please submit your proposal to the Editorial Board or to Francesca Bonadei, Executive Editor Mathematics, Statistics, and Engineering: francesca.bonadei@springer.com.

basque center for applied **mathematics**

Vadim Utkin · Alex Poznyak ·
Yury V. Orlov · Andrey Polyakov

Road Map for Sliding Mode Control Design

Vadim Utkin
Ohio State University
Columbus, OH, USA

Yury V. Orlov
Electronics and Telecommunications
CICESE Research Center
Ensenada, Baja California, Mexico

Alex Poznyak
Control Automatico Dpto
Cinvestav-Inst Politécnico Nacional
Mexico, Distrito Federal, Mexico

Andrey Polyakov
NON-A team of Inria in Lille
Inria Lille-Nord Europe
Villeneuve-d'Ascq, France

ISSN 2191-8198   ISSN 2191-8201 (electronic)
SpringerBriefs in Mathematics
ISBN 978-3-030-41708-6   ISBN 978-3-030-41709-3 (eBook)
https://doi.org/10.1007/978-3-030-41709-3

This Springer imprint is published by the registered company Springer Nature Switzerland AG
The registered company address is: Gewerbestrasse 11, 6330 Cham, Switzerland

Preface

This book reviews the basic ideas of sliding mode control (SMC) theory and demonstrates the deep interconnection of its different elements. The scope of the book is broad, encompassing the main design principles applied to a wide range of problems in control, observation, and information processing. The book is oriented toward engineers and theoreticians working in any area of control theory and applications.

In control systems, SMC is a nonlinear control method that alters the dynamics of a nonlinear system by application of a discontinuous control signal that forces the system to slide along manifolds with desired motions in the system state space. The state feedback control law is not a continuous function of time. Instead, such a system can switch from one continuous structure to another based on the current position in the state space. This explains why SMC stems from the so-called variable structure control method. In the context of modern control theory, any variable structure system, like a system under SMC, may be viewed as a special case of dynamical systems, supplied by switching logic.

Over the course of the development of the sliding modes theory during the past half-century, the range of both control problems and design methods to solve them has expanded significantly. Despite their great diversity, the basic principles of the SMC theory have remained unchanged since the creation of the necessary mathematical tools. The main control methods in sliding modes allow decomposition of the original system into independent subsystems of lower dimension and identification of the class of systems with sliding modes that are insensitive to various types of uncertainty in the description of the control object. These properties predetermine the success of application of the sliding modes theory for a wide class of dynamic objects: finite-dimensional models with scalar and vector inputs and outputs, models with delay, infinite-dimensional (described by partial differential equations) models, discrete-time models, stochastic models, etc.

The synthesis control procedure within the SMS is divided into two stages. In the first stage, it is necessary to choose the manifold in the state space of the system with the desired trajectories. This problem is equivalent to the design of a control in the systems of reduced dimension, for which any appropriate method of control theory is applicable. At the second stage, the control is selected to ensure the existence of a sliding mode on this preselected manifold. This is also a problem of reduced order with the same dimensions of the state and control vectors.

The task of providing a sliding mode is equivalent to the problem of stability of the zero solution of a nonlinear differential equation with a discontinuous right-hand side. The book provides a solution to this problem for systems of a fairly general form based on the Lyapunov function method. In particular, the described method allows us to reduce the well-known problems of root placement and quadratic optimization to solve similar problems of reduced dimension. All these properties are preserved when designing state observers for dynamical systems.

The main obstacle to the implementation of sliding modes is a phenomenon called chattering, which is caused by the mismatch between the ideal model and the real process. Within the power electronics literature, it is frequently referred to as ripple. In order to effectively suppress the chattering effect, the method of adapting the amplitude of the discontinuous signal and the harmonic cancellation method are used. These methods are applicable to both scalar and vector control systems. Note that in systems with scalar control, the sliding manifold is usually a surface in the state space. The high-order sliding mode method assumes that the sliding manifold can have a smaller dimension. Accordingly, the sliding mode equation has a smaller dimension. However, the problem of enforcing the sliding mode becomes more difficult as the dimension of control is less than the dimension of the system to be stabilized.

Methods developed for time-continuous systems have proved unsuitable for discrete-time systems because discontinuous control always leads to chattering even within the framework of an ideal model. It turns out that the main property of the SMC, namely, the achievement of the desired manifold in a finite time and the further motion in it, can be saved by using a control as a continuous function of the system state vector. This property has made it possible to develop methods for the synthesis of SMC for discrete systems.

SMC synthesis admits the presence of free parameters (such as the amplitude of discontinuous control and parameters of sliding manifold equations), which can be used to significantly improve the accuracy and dynamics of control processes. Such methods for designing adaptive control laws are applicable in systems with sliding modes.

The use of SMC for infinite-dimensional systems required a revision of the basic concepts of the theory. Questions of the mathematical description of solutions of partial differential equations with a discontinuous right-hand side had hardly been considered in the literature. Component-wise synthesis methods, developed for finite-dimensional systems, proved to be unacceptable. One of the chapters in this book is devoted to the solution of these problems, where distributed thermal processes and flexible mechanical systems are considered as applications.

An attempt to extend the SMC theory to the class of systems with random perturbations requires the use of the modern theory of stochastic systems. These studies are at an initial stage, and the first results are also reflected in this book.

Finally, further perspectives of the research are outlined in the concluding chapter.

The book reflects the consensus view of the authors regarding the current status of SMC theory, as it emerged from preceding discussions over a long period. We would like to underline that the authors are affiliated with different research centers and that their positions at the beginning of the discussions were far from being identical, as they are now. Finally, we would also like to point out that the book reflects the views of four generations of colleagues with rich experience in the area, the age difference between these generations being 10 years or more in each instance.

Columbus, USA   Vadim Utkin
Mexico, Mexico   Alex Poznyak
Ensenada, Mexico   Yury V. Orlov
Villeneuve-d'Ascq, France   Andrey Polyakov

Contents

About the Authors

Vadim Utkin graduated from Moscow Power Institute (Dipl. Eng.) and received Ph.D. and Doctor of Science degrees from the Institute of Control Sciences (Moscow, Russia). He worked at the Institute of Control Sciences from 1960 to 1994, and in 1973 was appointed as Head of the Discontinuous Control Systems Laboratory. Currently, he is a Professor at Ohio State University. He is one of the originators of the concepts of variable structure systems and sliding mode control. His application interests are control of power converters and electric drives, robotics, and automotive control. He is the author or co-author of five books and 350 papers.

Alex Poznyak graduated from Moscow Physical Technical Institute (MPhTI) in 1970. He earned Ph.D. and Doctor of Science degrees from the Institute of Control Sciences of the Russian Academy of Sciences in 1978 and 1989, respectively. From 1973 to 1993, he served first as a researcher and then as leading researcher at this institute, before accepting a post as Full Professor (3-F) at the Center for Research and Advanced Studies of the National Polytechnic Institute (CINVESTAV-IPN) in Mexico, where for 8 years he was Head of the Automatic Control Department. He has been the supervisor for 43 Ph.D. theses. He has published more than 240 papers in different international journals and 14 books. His areas of interest are robust nonlinear deterministic and stochastic control, identification theory, Markov processes, and game theory with economic applications.

Yury V. Orlov is a Professor in the Electronics and Telecommunication Department, Scientific Research and Advanced Studies Center of Ensenada, Mexico. His research interests lie in the analysis and synthesis of discontinuous as well as time-delay and distributed-parameter systems. He has authored or co-authored about 250 journal and conference papers in the above areas as well as five monographs. He is an Associate Editor of IEEE Transactions on Control Systems Technology, of the International Journal of Robust and Nonlinear Control, and of the IMA Journal of Mathematical Control and Information.

Andrey Polyakov received his Ph.D. in Systems Analysis and Control from Voronezh State University in 2005. Until 2010, he was an Associate Professor with this university. In 2007–2008, he worked at CINVESTAV-IPN in Mexico. From 2010 to 2013, he was a Lead Researcher of the Institute of Control Sciences of the Russian Academy of Sciences, and he then joined INRIA, Lille, France. His main research interests include robust and nonlinear control. He has authored or co-authored more than 150 papers.

Chapter 1
Introduction

Abstract The initial ideas for designing systems with a sliding mode were formulated half a century ago for dynamic models in canonical space with scalar input and output. The interest in the behavior analysis in the space of the output variable and its time derivatives was explained by the property of invariance with respect to perturbations and parametric variations. This invariance property is lost for arbitrary space in systems with vector input and output, which predetermined the need for new mathematical methods for the analysis of such systems and design of control methods with sliding modes. The set of these questions is the content of this book.

Keywords Sliding mode · Canonical space · Perturbations and parametric variations

The multi-branched tree of *Sliding Mode Control* (SMC) theory was seeded more than 50 years ago. At the very beginning, only SISO systems in the canonical space were studied. The interest in the space of output signal and its time derivatives can be explained easily: the sliding mode (SM) equation depends on neither parameter variations nor external disturbances (see Fig. 1.1).

The example in Fig. 1.1 demonstrated that the motion in sliding mode is of a reduced order and depends on the switching line equation only. It looks attractive to utilize these properties for control at the presence of disturbances. We compare the second-order systems with SMC and PID controller. As shown in Fig. 1.2, after a finite time interval, the sliding mode occurs (since the distance to the switching line s becomes equal to zero identically); the output does not depend on disturbance and tends to zero exponentially. Parameters of PID controller are selected such that the transient time for both processes are the same. The output of the system with PID controller cannot be reduced to zero (Fig. 1.3). An integral component in PID controller is able to reject constant disturbances only. The system with SMC is equivalent to the system with a high feedback gain, implemented by finite control actions. Therefore, an integral component is not needed for SMC, and steady-state error is equal to zero identically in contrast to systems with a linear PID controller.

SMC design method *can be generalized* easily for systems of an arbitrary order in canonical form

$$x^{(n)} + a_n x^{(n-1)} + \cdots + a_1 x = u + d,$$

© The Author(s), under exclusive license to Springer Nature Switzerland AG 2020
V. Utkin et al., *Road Map for Sliding Mode Control Design*,
SpringerBriefs in Mathematics,
https://doi.org/10.1007/978-3-030-41709-3_1

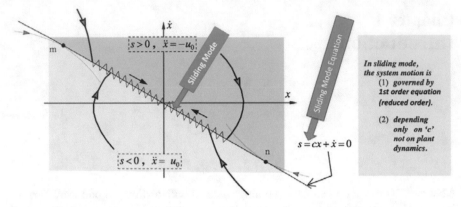

Fig. 1.1 Sliding mode in the second-order system

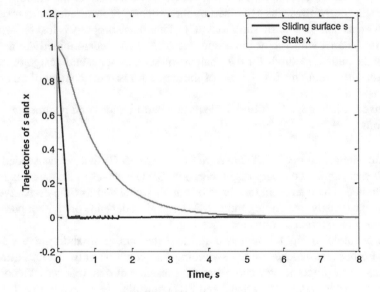

Fig. 1.2 $\ddot{x} = u + \sin(3t)$, SMC: $u = -3\,\mathrm{sign}\,(s)$, $s = \dot{x} + x$

$$u = -M\,\mathrm{sign}(s), \quad s = \sum_{i=1}^{n} C_i x^{(i-1)}, \ C_n = 1.$$

Similar to the second-order system, sliding mode is enforced on the switching plane $s = 0$, governed by equation

$$\sum_{i=1}^{n} C_i x^{(i-1)} = 0.$$

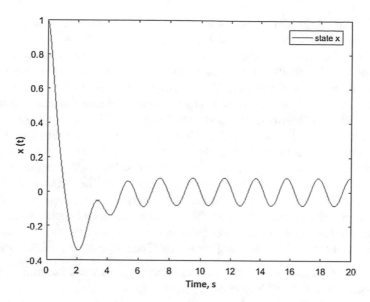

Fig. 1.3 $\ddot{x} = u + \sin(3t)$, PID controller: $u(s) = -(7 + 3.5/s + 4.5s)x(s)$

This motion depends on neither parameters a_i nor disturbance d. SMC in canonical space was studied practically in all chapters of [1]. SMC in the canonical space needs high-order time derivatives, and their ideal implementation is hardly possible. As a rule, an output of any real differentiator is approximation of operator d/dt, which leads to undesirable high-frequency oscillations, called the chattering. Chattering is recognized as the main obstacle for implementation of SMC.

Sliding mode in the canonical space does not depend on uncertainties. Unfortunately, this property is not inherent in sliding modes in an arbitrary state space. The equation of a switching plane is not the equation of the sliding mode simultaneously. The question of identifying a sliding mode equation in an arbitrary state space is not trivial and needs special study. All above mentioned problems have been studied in further (after [1]) publications and discussed in this book.

Since then the range of design methods and control problem statements have been increased considerably. Now the scope of research embraces the design in an arbitrary state space, MIMO systems, the observation problems, and mathematical methods of analysis. Publication of survey papers in the 70–80s [2, 3], embracing the scope of research in the area, stimulated proliferation of SMC methodology into research programs of control theorists and engineers in many countries. As a result the scientific arsenal, accumulated in the theory of SMC during more than 50 years, offers a wide range of both problem statements and methods for their solutions. So wide that any attempt to prepare a "State-of-the-Art" survey paper would hardly be successful. Of course, excellent surveys were published about different aspects on SMC (chattering problems, adaptive control, high-order sliding mode, applications). The list can be complemented by control in infinite-dimensional and stochastic systems, generalization for discrete-time systems and integral control.

We are realistic and far from the idea of preparing a survey embracing the results of numerous publications in those new areas. Our objective is more modest; we try to describe the basic ideas of the SMC methodology and their development in a historical perspective, not to describe only but to demonstrate deep interconnection between so-called "conventional" and "new" methods.[1] In the course of the SMC history, the original ideas were modified in the context of new control problems, but none of them was recognized as obsolete and replaced by a totally new idea. The tracks of the main concepts formulated in "childhood" of sliding mode control can be readily observed in any modern research. In conclusion, the further developments are outlined. The recent results are further development of the "conventional" SMC theory rather than an attempt to replace it. The accent in all sections is made on revealing the reasons why new developments were needed and why their roots can be found easily in the "conventional" methods. The purposes of the book are

- to inform the readers on a wide range of control problems successfully solved in the framework of the SMC theory;
- to demonstrate that the background of all methods is the same: selection of a reduced-order differential equation with desired solutions and implementation of the selected motion by enforcing sliding mode;
- to demonstrate the potential and advantages of the developed control methods and outline the areas of their efficient applications.

The book is oriented toward a wide range of engineers and theoreticians working in any area of the control theory and applications. All chapters of the book are prepared in this context. The scope embraces control of finite- and infinite-dimensional processes with continuous- and discrete-time control, chattering suppression problems, and new methods of adaptation applicable for systems with sliding motions only. Special mathematical methods are needed for all listed control tasks. The methods constitute the initial chapters and include the definition of the multidimensional SM, the derivation of the differential equations of those motions, and the existing conditions. The book also discusses areas of further research.

In our view, it is better to avoid separation of past and current times but establish bridges between them.

---

[1] We follow the terminology: "conventional" means the first-order sliding mode, "new" means the high-order sliding mode. The term "conventional SMC" appeared in Shtessel et al. (2014), for the first time. The details will be discussed in the subsequent chapters.

References

1. Emelyanov, S. (ed.): Theory of Variable Structure Control Systems (in Russian). Nauka (1970)
2. Utkin, V.: Variable structure systems with slidnng mode control. IEEE Trans. Autom. Control **22**(2), 212–221 (1977)
3. DeCarlo, R., Zak, S., Matthews, G.: Variable structure control of nonlinear multivariable systems: a tutorial. Proc. IEEE **76**(3), 212–232 (1988)

Chapter 2
Mathematical Methods

Abstract This chapter presents the definition of sliding mode and discusses the conditions of the existence and the uniqueness of the solution of the arising differential inclusion. The definition of the Filippov solution is introduced and discussed. The equivalent control method (ECM) is presented as an alternative procedure for deriving the sliding mode equation. Several types (Filippov, Utkin) of the regularization methods are discussed and illustrated by the class of affine systems.

Keywords Sliding mode · Filippov solution · Regularization methods

2.1 Definition of Sliding Mode

State space representation is the conventional framework for analysis and design of sliding mode control systems. A plant model in this case is described by some ordinary differential equations (ODEs)

$$\dot{x}(t) = f(t, x(t), u(t, x(t))), \quad t \in \mathbb{R}, \tag{2.1}$$

where $x(t) \in R^n$ is the vector of system states, $u(t) \in R^m$ is the vector of control inputs, and f is a function. For simplicity, we assume that f is continuous, but the control u, as a function of time and state, can be discontinuous in its arguments but locally bounded. Later on, we discuss different definitions of a solution to the discontinuous ODE (2.1). Here, we just assume that in a domain $G \subseteq R^{n+1}$ the closed-loop system (2.1) has absolutely continuous solutions, which are differentiable almost everywhere and the derivatives are locally bounded.

The main idea of sliding mode control consists of enforcing the trajectories of closed-loop system to belong to a prescribed manifold (*sliding manifold*). The manifold has to be selected in advance in order to fulfill the required performance of the closed-loop system. Next, the control law has to be designed to enforce the sliding mode (see details in Chap. 4).

To introduce a definition of sliding mode for a finite-dimensional model (2.1), we try to find a minimal set of properties characterizing it. The finite-time convergence to a sliding set is a specific feature of the finite-dimensional sliding mode systems [1–4]. That is impossible for ODEs with Lipschitzian right-hand sides. The discontinuity of the sliding mode control provides some robustness properties for the closed-loop system such as invariance with respect to the matched perturbations [5].

Definition 2.1 Let S be a r-dimensional manifold in the state space, $r < n$. An open non-empty subset $D \subseteq S$ is said to be *sliding domain* if it is locally uniformly finite-time attractive, that is, for any $y \in D$ there exists $\varepsilon > 0$ such that, if

$$x(t_0) = x_0, \|x_0 - y\| < \varepsilon,$$

then $T(x_0)$ can be found such that

$$x(T(x_0)) \in D \text{ and } \lim_{x_0 \to y} T(x_0) = 0,$$

where x is a solution of the closed-loop system (2.1) with the initial condition $x(t_0) = x_0$.

If $D = S$, then D is called *sliding manifold* (Fig. 2.1). The conventional description of sliding manifold is
$$S = \{x \in R^n : s(x) = 0\},$$

where s is a smooth function that maps R^n to R^r. For ease of reference, it is often written as $s(x) = 0$ or simply $s = 0$. Our definition based on finite-time convergence cannot be called new. Similar definition was introduced 25 years ago in [1], see also Chap. 4 in [3].

The definition does not imply discontinuities in control. As a result, the term "sliding mode" proved to be applicable for discrete-time systems. For several decades, starting from 50s the term "sliding mode" was associated with discontinuities in motion equations and with trajectories in the discontinuities surface. Attempts to extend it for other classes of systems based on this view were not successful. For example, discontinuities in discrete-time systems resulted in chattering. An alternative approach is to make behavior of discrete-time systems similar to continuous-time ones so that an integral manifold exists with finite reaching time. This approach proved to be appropriate for discrete-time systems, when sliding mode can appear, if right-hand side in motion equation is continuous (similar to dead-beat control [6]), as well as for continuous-time systems with continuous non-Lipschitz functions. In nutshell, sliding mode does not imply discontinuities always. Another interpretation of the systems with sliding modes was offered in 1953 by Pontryagin. He said, a system with "glued" trajectories does not have an inverse shift operator, and it is even not a dynamic system. We follow this concept and in terms of our definition, a sliding mode exists in the continuous-time system with a square root in the equation (see [7]). The finite-time stable scalar system

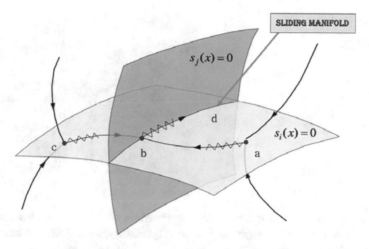

Fig. 2.1 Sliding mode in discontinuity surfaces and their intersections

$$\dot{x}(t) = -\sqrt{|x(t)|}\,\text{sign}(x(t))$$

is topologically equivalent to the system

$$\dot{y}(t) = -0.5\,\text{sign}(y(t)),$$

obtained by means of change of the coordinate

$$y = \sqrt{|x|}\,\text{sign}(x).$$

After a finite time, y is identically nullified and as a result, the state x is as well in spite of the continuous right-hand side in the original equation. Indeed, the coordinate transformation is homeomorphism (continuously invertible) on R; moreover, it is diffeomorphism (smooth) on $R\backslash\{0\}$. No need to say, that continuous non-smooth control resulting in SM is not applicable for rejection of additive disturbances, since it is equal to zero on the desired sliding surface.

Minor remark: *Excluding finite-time attractiveness property, we would certainly admit that sliding mode exists in linear system. Widely used in literature inequality*

$$s\dot{s} < 0 \text{ for } s \neq 0$$

as the condition for sliding mode to exist does not imply finite-time attractiveness property, as in the above definition. Accepting this property, we should admit that sliding mode exists on the line $s = 0$ in the linear second-order system

$$\ddot{x} = x, \quad s = \dot{x} - x,$$

Fig. 2.2 Illustration of the
situation with no sliding
mode in origin

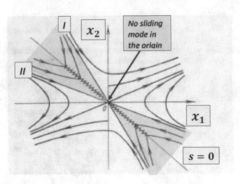

since $\dot{s} = -s$ and $s\dot{s} < 0$ if $s \neq 0$. One more example illustrating that condition
$s\dot{s} < 0$, if $s \neq 0$ is not sufficient for existence of sliding mode. Sliding mode occurs
on the switching line $s = 0$ with

$$s = \dot{x} + x$$

after finite time for any initial conditions from the open blue sector in the system (see
Fig. 2.2, where $x_1 = x$ and $x_2 = \dot{x}$).

$$\ddot{x} = \begin{cases} 0.5x & \text{if } x \cdot s > 0 \\ 1.5x & \text{if } x \cdot s < 0 \end{cases}.$$

Inequality $s\dot{s} < 0$ holds and s tends to zero for initial conditions on lines I and II;
however, the sliding mode does not exist in the origin; there are diverging trajectories
in any its vicinity.

For systems with scalar control undergoing discontinuities on surface $s(x) = 0$,
the sufficient conditions for sliding mode to exist [8, Chap. 2] are

$$\lim_{s \to 0^+} \dot{s} < 0 \quad \text{and} \quad \lim_{s \to 0^-} \dot{s} > 0. \tag{2.2}$$

For example, the sector mn (Fig. 2.3) is a sliding domain for the second-order system

$$\begin{rcases} \ddot{x} = u, \quad u = -\text{sign}(s), \\ s = \dot{x} + cx, c > 0. \end{rcases} \tag{2.3}$$

Note that $x(t)$ tends to zero as a *stable solution* (see Fig. 2.3) to equation

$$s = \dot{x} + cx = 0,$$

and the state $(x, \dot{x})$ never reaches the boundary point m or n, should sliding mode
occurs. Sliding mode exists on segment mn of switching line

$$\dot{x} + cx = 0, \quad c < 0$$

Fig. 2.3 Stable solution

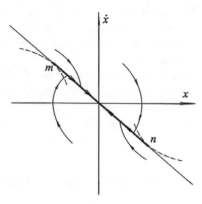

Fig. 2.4 Unstable solution

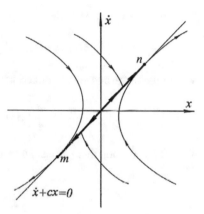

(see Fig. 2.4), but the motion is *unstable*. It is important to note that the statements about stability are based on heuristic arguments. The solution

$$x(t) = e^{-c(t-t^*)}x(t^*)$$

is found from equation

$$\dot{x} + cx = 0.$$

Formally, it should satisfy equation

$$\ddot{x} = u$$

or

$$x(t^*)c^2 e^{-c(t-t^*)} = -\text{sign}(0),$$

where $t^* \in R$ is a time instant, when the trajectory reaches the sliding set. The questions arise why the latter equality holds and what meaning is behind $\text{sign}(0)$.

The questions touch upon the fundamental problem—what the solution is, since the conventional existence-uniqueness theorems [9] are not applicable for differential equations with discontinuous right-hand sides. The above problem is treated in the next section.

It is worth noticing that Definition 2.1 does not specify a class of dynamical systems. It is applicable even for continuous non-Lipschitz ODEs and the so-called high-order sliding mode (HOSM) systems [10], [4, Chap. 6] which are discussed in Chap. 7 of this book.

2.2 Solution Existence and Uniqueness Problem

The simplest example of the sliding mode control system is

$$\dot{x}(t) = u(t) + d(t), \quad u(t) = -\operatorname{sign}(x(t)),$$

where the disturbance is bounded $|d(t)| \leq d_0 < 1$. Since

$$\frac{d}{dt}|x(t)| \leq -(1 - d_0) < 0$$

for $x(t) \neq 0$, then there exists an instant of time

$$T(x(t_0)) \leq t_0 + \frac{|x(t_0)|}{1 - d_0}$$

such that

$$|x(t)| \to 0 \text{ as } t \to T(x_0) + 0^-.$$

The mathematical obstruction appears if we try to define a solution for $t > T(x(t_0))$ because classical concepts of solutions imply the existence of a Lipschitz constant [9] and do not allow the right-hand side of ordinary differential equation to be discontinuous. It is reasonable to assume that $x(t) = 0$ for $t > T(x(t_0))$. As in the example of Sect. 2.1, the question arises again, why equality

$$-\operatorname{sign}(0) = d(t)$$

holds. The unanswered question serves as an example, illustrating, why the conventional theory is not applicable to discontinuous systems and why new mathematical methods for discontinuous systems should be developed. The methods should answer the fundamental questions of the classical theory of differential equations if a solution exists at points of discontinuities and if it exists, whether the solution is unique or not.

The analysis is far from being trivial for multidimensional and nonlinear cases. It may lead to solutions, which contradict engineering intuition.

Fig. 2.5 Real sliding mode for relay with hysteresis

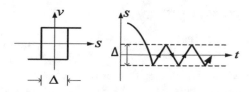

Fig. 2.6 Phase shift for two relay controls with hysteresis

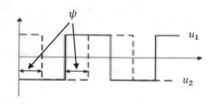

Let us consider the system

$$\left. \begin{aligned} \dot{x}_1(t) &= u_1(t), \quad u_1(t) = -\text{sign}(x_1(t)), \\ \dot{x}_2(t) &= u_2(t), \quad u_2(t) = -\text{sign}(x_2(t)), \\ \dot{x}_3(t) &= u_1(t)u_2(t). \end{aligned} \right\} \tag{2.4}$$

The state variables in the first two equations and their time derivatives have opposite signs; therefore, the sliding mode exists with $x_1(t) = 0$ and $x_2(t) = 0$. Assuming $u_1(t) = 0$ and $u_2(t) = 0$ yields

$$\dot{x}_3(t) = 0 \text{ or } x_3(t) = \text{const.}$$

If u_1 and u_2 are relay functions with hysteresis Δ (Fig. 2.5, $v = u_1$ or u_2, $s = x_1$ or x_2), the average value of the product $u_1 u_2$ can take any value between $+1$ and -1 depending on the phase shift ψ (Fig. 2.6). The high-frequency component in the solution to the third equation is filtered out with Δ tending to zero. Therefore, x_3 is the solution to

$$\dot{x}_3(t) = A, \quad -1 \le A \le 1, \quad x_3 = At + x_3(0)$$

(not constant as stated before).

If the equations for x_1 and x_2 are studied separately, we obtain the sliding motion on the surfaces

$$S_1 = \{x = (x_1, x_2, x_3)^T \in \mathbb{R}^2 : x_1 = 0\}$$
$$S_2 = \{x = (x_1, x_2, x_3)^T \in \mathbb{R}^2 : x_2 = 0\},$$

respectively. One can define $u_i(t) = 0$ for $t > T_i(x(t_0))$, where $T_i(x(t_0))$ is an instant of time when the trajectory reaches the surface S_i, $i = 1, 2$. In this case, $\dot{x}_3(t) = 0$ for $t \ge T_i(x(t_0))$.

The proposed procedure provides the well-defined unique continuous solution to the considered system. However, in practice, the relay control is realized with some defects, for example, delay and/or hysteresis. If the control inputs are replaced with

$$u_i^h(t) = -\text{sign}(x_i(t - h)),$$

where $h > 0$, then the corresponding solution $x_i^h(t), i = 1, 2, 3$ of the closed-loop system with $x_i^h(t) = 0$ for $t \in [-h, 0)$ is well defined in the sense of Carathéodory. Moreover, it can be shown that $x_i^h(t) \to x_i(t)$ as $h \to 0$ uniformly on t for $i = 1, 2$, but $x_3^h(t)$ may not converge to $x_3(t)$. Indeed, if

$$x_1^h(t_0) = \pm x_2^h(t_0) \neq 0,$$

then

$$u_1^h(t) = \pm u_2^h(t),$$
$$\dot{x}_3^h(t) = u_1^h(t)u_2^h(t) = \pm 1$$

for almost all $t > 0$. The last identity contradicts to

$$\dot{x}_3(t) = 0, t > T^* = \max\{T_1(x(t_0)), T_2(x(t_0))\}$$

derived in delay-free case. It can be shown that for any $h > 0$, the initial conditions can be selected in such a way that the averaged value of time derivative $\frac{1}{4h}\int_t^{t+4h} \dot{x}_3^h(\tau)d\tau$ will take any value in $(-1, 1)$ for $t > T^*$.

2.3 Filippov Solution

Filippov method [11, Chap. 2] is the classical tool to overcome the mathematical obstructions of the discontinuous ODE. The main idea is to embed the ill-posed ODE system into a well-posed differential inclusion, which coincides with ODE in continuity domain. Let a time-varying feedback law u in (2.1) be a discontinuous function of state variables. The Filippov method of solution continuation at any point of discontinuity set $N \subseteq R^n$ of measure zero implies that the right-hand side of (2.1) belongs to the convex hull of all vectors in its vicinity. It is important to note that Filippov's method has a very simple interpretation in the time domain. Let the right-hand side of (2.1) take one of k possible values $\tilde{f}_1, \ldots, \tilde{f}_k$ in the vicinity of some point (t_0, x_0) in the state space, and a time interval Δt consists of non-overlapping k subsets

$$\Delta t_1, \Delta t_2, \ldots, \Delta t_k : \Delta t = \sum_{i=1}^{k} \Delta t_k$$

with values of right-hand sides $\tilde{f}_1, \ldots, \tilde{f}_k$ correspondingly. Then, the averaged speed in this point satisfies

$$\dot{x}(t) = \lim_{\Delta t \to 0} \frac{1}{\Delta t} \sum_{i=1}^{k} \Delta t_i \, \tilde{f}_i(t, x(t)), \quad \alpha_i = \frac{\Delta t_i}{\Delta t}, \quad \sum_{i=1}^{k} \alpha_i = 1.$$

The right-hand side is nothing else but the convex hull of $\tilde{f}_1, \ldots, \tilde{f}_k$. Therefore, α_i can be viewed as a specific time of the mode $\tilde{f}_i$. The Filippov procedure considers the following differential inclusion:

$$\dot{x} \in F(t, x(t)), \quad t \in \mathbb{R},$$

provided that the set-valued function $F : \mathbb{R}^{n+1} \to 2^{\mathbb{R}^n}$ is defined as

$$F(t, x) = \bigcap_{\varepsilon > 0} \bigcap_{N : \mu(N) = 0} \mathrm{co} \tilde{f}(t, x + \varepsilon B \backslash N), \qquad (2.5)$$

where

- $\tilde{f}(t, x) = f(t, x, u(t, x))$ is a right-hand side of the closed-loop system (2.1),
- $B = \{x \in R^n : \|x\| \le 1\}$ is the unit ball in R^n,
- $\mathrm{co}(M)$ is the *convex* hull of a set M,
- and the identity $\mu(N) = 0$ means that the set N is of measure zero.

The Filippov procedure is applicable for any locally measurable (possibly nowhere continuous) function $\tilde{f}$. If it is known *a priori* that $\tilde{f}$ is discontinuous only on a set N of the measure zero, then the second intersection can be excluded from (2.5).

Definition 2.2 An absolutely continuous function x is a solution of (2.1), if almost everywhere it satisfies the inclusion

$$\left. \begin{array}{l} \dot{x}(t) \in F(t, x(t)), \qquad F(t, x) = \lim_{\varepsilon \to 0} F_\varepsilon(t, x), \\ F_\varepsilon(t, x) = \mathrm{Conv}\{f(t, x, u(t, \{x + \varepsilon B\} \backslash N))\}, \end{array} \right\} \qquad (2.6)$$

where B is the ball centered in the origin of the unit radius, $N \subseteq R^n$ is the discontinuity set of the measure zero, and the limit is defined by the Hausdorff distance,[1] $d(F(t, x), F_\varepsilon(t, x)) \to 0$ as $\varepsilon \to 0$.

Exclusion of a set of measure zero in (2.6) is the core condition when deriving the sliding mode equation. Any attempts to assign the right-hand side on discontinuity

---

[1]The Hausdorff distance between two non-empty sets $S_1, S_2 \subseteq R^n$ is a number

$$d(S_1, S_2) = \max\{d_1(S_1, S_2), d_2(S_1, S_2)\},$$

where

$$d_1(S_1, S_2) = \max_{x_1 \in S_1} \min_{x_2 \in S_2} \|x_1 - x_2\|$$

and

$$d_2(S_1, S_2) = \max_{x_2 \in S_2} \min_{x_1 \in S_1} \|x_1 - x_2\|$$

manifold beyond the context of the system equation can meet contradictions. In many publications

$$\text{sign}(\rho) = \begin{cases} 1 & \text{if } \rho > 0, \\ -1 & \text{if } \rho < 0, \\ 0 & \text{if } \rho = 0, \end{cases}$$

or

$$\text{sign}(\rho) = \begin{cases} 1 & \text{if } \rho \geq 0, \\ -1 & \text{if } \rho < 0, \end{cases}$$

which is **not true** for the value of $\text{sign}(0)$ in real system. This fact is demonstrated in the explanation for the system with friction. The friction force cannot be assigned, if the external force is unknown. For example, in the mechanical system

$$m\ddot{x} = g_{fr}(t) - g(t), t > 0,$$

where $x(t) \in R$, the Coulomb friction

$$g_{fr}(t) = -F_0 \text{sign}(\dot{x}(t))$$

undergoes discontinuities on surface $\dot{x} = 0$, and sliding mode occurs on this surface if $F_0 > |g(t)|$. It is evident that $g_{fr}(t) = g(t)$, and this time function does not coincide with an arbitrary assigned function (beyond the system context).

Let us write down sliding mode equation following the Filippov method for the autonomous system (2.1) with scalar control, that is, discontinuous on a smooth surface

$$S = \{x \in R^n : s(x) = 0, \quad s \in C^1\},$$

which separates R^n on two domains with

$$s(x) > 0, \quad f(t, x, u(x)) = f^+(t, x)$$

and with

$$s(x) < 0, \quad f(t, x, u(x)) = f^-(t, x),$$

where f^+ and $f^-(t, x)$ are some functions continuous on R^{n+1}.

For $x \in S$, the convex hull $F(x)$ is the segment

$$\mu f^+(t, x) + (1 - \mu)f^-(t, x), 0 \leq \mu \leq 1,$$

connecting vectors $f^+(t, x)$ and $f^-(t, x)$. The state velocity vector in sliding mode should be in the plane tangential to the surface S at a point $x \in S$. If the existence con-

Fig. 2.7 Geometrical
illustration of Filippov
definition

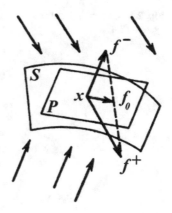

ditions (2.2) hold, then this segment crosses the plane (see Fig. 2.7). The intersection
point defines the right-hand side in the sliding mode equation in surface $s(x) = 0$

$$\dot{x}(t) = f_0(t, x(t)),$$

where

$$f_0(t, x) = \mu f^+(t, x) + (1 - \mu) f^-(t, x)$$

and the parameter μ can be found from condition

$$\nabla s \perp [\mu f^+ + (1 - \mu) f^-].$$

Substituting the solution into the formula for f_0 results in

$$f_0 = \frac{(\nabla^T s \, f^-) f^+ + (\nabla^T s \, f^+) f^-}{\nabla^T s \, (f^+ - f^-)}.$$

If $\nabla s(x)$ is not orthogonal to $\mu \tilde{f}^-(t, x) + (1 - \mu) \tilde{f}^+(t, x)$ for every $\mu \in [0, 1]$, then
any trajectory of (2.6) comes through the surface resulting in an isolated "switching"
of the right-hand side of (2.1). Therefore, for the system with scalar control and two
possible vectors f^+ and f^-, the right-hand side of the sliding mode equation should
belong to the convex hull of these two vectors.

Now let us consider again the example (2.4). From the first impression, $u_1(t)$ and
$u_2(t)$ should be equal to zero since $x_1(t)$ and $x_2(t)$ are equal to zero identically in
sliding mode and as a result

$$\dot{x}_3(t) = 0, \, x_3(t) = \text{const}.$$

However, both of them are the same pulse trains and depending on the phase shift
the average value of their product can take any value between -1 and $+1$. So, the
implementation of the control inputs by a relay with hysteresis showed that

Fig. 2.8 Convex hull for the
system (2.4)

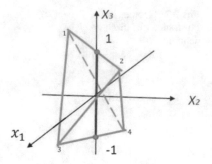

$$\dot{x}_3(t) = A, \quad -1 \le A \le 1, \quad x_3(t) = x_3(0) + At.$$

Validity of this equation can also be confirmed by Filippov's method. Four possible state speed vectors correspond to four combinations of two control functions. The tetrahedron is their convex hull (Fig. 2.8), and its intersection with manifold $x_1 = x_2 = 0$ (axis x_3) is segment $[-1, 1]$, which defines the same all possible sliding mode equations

$$\dot{x}_3(t) = A, \quad -1 \le A \le 1.$$

2.4 Equivalent Control Method

An alternative procedure for deriving the sliding mode equation $\dot{x}(t) = f_0(t, x(t))$ is presented in this section for control system (2.1) under the assumption that f is continuous on R^{n+m+1}, but each component u_i of control $u = (u_1, \ldots, u_m)^T$ is discontinuous only on a smooth surface $s_i(x) = 0$. Assume that sliding mode occurs in the intersection of k switching surfaces $s^k(x) = 0$, where $s^k(x)$ is k-dimensional vector consisting of k arbitrary components of the set $\{s_1(x), \ldots, s_m(x)\}$. Vector u^k consists of corresponding components (of course $u^k = u$ if $k = m$). Vector $f_0(t, x)$ should belong to the space tangential to

$$s^k(x) = 0, \text{ or } \nabla^T s(x) f_0(t, x) = 0,$$

where $\nabla^T s(x)$ is $k \times n$ matrix consisting of gradients of components of s^k as rows. It looks natural to replace the discontinuous control in (2.1) by a continuous state function, such that the condition $\nabla^T s^k(x) f_0(t, x) = 0$ holds. Consequently, a solution to equation $\nabla^T s^k(x) f(t, x, u) = 0$ with respect to u should exist. The solution u_{eq}^k called *equivalent control* [1, Chap. 2] is substituted into (2.1) to get the sliding mode equation. The method of deriving the sliding mode equation is called *equivalent control method*. Application of the equivalent control and Filippov's methods to Eq. (2.1) with the scalar control

Fig. 2.9 Different velocity vectors: Filippov's and equivalent control methods

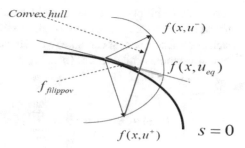

Fig. 2.10 Different velocity vectors: Filippov's and equivalent control methods **(b)**

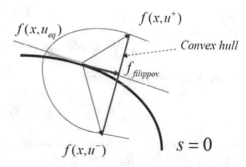

$$u(t,x) = \begin{cases} u^+(t,x) \\ u^-(t,x) \end{cases} \quad \text{and} \quad f(t,x) = \begin{cases} f(t,x,u^+(t,x)) \\ f(t,x,u^-(t,x)) \end{cases}$$

demonstrates that, in general, they lead to different sliding mode equations (see Figs. 2.9, 2.10). Indeed as follows from Filippov's method, the right-hand side of the sliding mode equation is defined by the intersection point of the tangential plane and the straight line, connecting the ends of f^+ and f^- (convex hull). The sliding mode equation for the equivalent control method is defined by the intersection point of the tangential plane and locus $f(t,x,u)$ with respect to u (vector $f(t,x,u_{eq})$ should be in the tangential plane).

Since both methods were postulated, therefore, the question "which of them is correct" cannot be answered. Another important engineering question "which sliding mode equation is in real-life systems" will be discussed in the next subsection.

It is interesting that for affine (linear with respect to control) systems of the form

$$\left. \begin{aligned} \dot{x}(t) &= f(t,x(t)) + B(t,x(t))u(t,x(t)), \quad x(t) \in R^n, \\ u_i(t,x) &= \begin{cases} u_i^+(t,x) & \text{if } s_i(x) > 0, \\ u_i^-(t,x) & \text{if } s_i(x) < 0, \end{cases} \\ u &= (u_1,\ldots,u_m)^\mathsf{T}, \quad s = (s_1,\ldots,s_m)^\mathsf{T}, \end{aligned} \right\} \tag{2.7}$$

the both methods lead to the same sliding mode equation [1, p. 35]

$$\left. \begin{aligned} \dot{x}(t) &= f_0(t,x(t)), \\ f_0(t,x) &= f(t,x) - B(t,x)(\nabla^\mathsf{T}s(x)B(t,x))^{-1}\nabla^\mathsf{T}s(x)f(t,x). \end{aligned} \right\} \tag{2.8}$$

According to Filippov's method, right-hand side f_0 of the sliding mode equation should belong to the convex hull of all possible vectors of the original system (2.7). Each control component can take two values; hence, the total number of all possible control vectors is equal to 2^m, $\{v^1, \ldots, v^{2^m}\}$. The intersection of convex hull of the right-hand side of (2.7)

$$\left\{ f(t, x) + B(t, x)z : z = \sum_{i=1}^{2^m} \mu_i v^i, \quad 0 \le \mu_i \le 1, \quad \sum_{i=1}^{2^m} \mu_i = 1 \right\}$$

with the tangential manifold to $s(x) = 0$ defines sliding mode equation. It follows that equation

$$\nabla^T s(x) f(t, x) + \nabla^T s(x) B(t, x)z = 0$$

holds. Substitution of its solution

$$z = -(\nabla^T s(x) B(t, x))^{-1} \nabla^T s(x) f(t, x)$$

into the equation for the convex hull results in the sliding mode equation. It coincides with Eq. (2.8) resulting from equivalent control method.

An alternative procedure can be presented for control system (2.1) under assumption that f is continuous on all arguments, but each component u_i of a feedback law

$$u : \mathbb{R}^{n+1} \to \mathbb{R}^m, \quad u(t, x) = (u_1(t, x), u_2(t, x), \ldots, u_m(t, x))^T$$

is discontinuous only on a surface

$$S_i = \{(t, x) \in \mathbb{R}^{n+1} : s_i(t, x) = 0\},$$

where $s_i : \mathbb{R}^{n+1} \to \mathbb{R}$ is a smooth function.

Let us construct the following multivalued function:

$$U(t, x) = \bigcap_{\varepsilon > 0} \bigcap_{\mu(N) = 0} \text{co } u (t, x + \varepsilon B \backslash N).$$

Definition 2.3 An absolutely continuous function $x : R \to R^n$ defined on some interval or segment I is a solution to (2.1) if there exists a locally measurable function $u_{eq} : R \to R^m$ such that

$$u_{eq}(t) \in U(t, x(t))$$

and

$$\dot{x}(t) = f(t, x(t), u_{eq}(t))$$

almost everywhere on I.

In [12, 13], the presented solution was called *the Utkin solution*, since it follows the main idea of *Equivalent Control Method* proposed by Utkin [1]. Obviously, for $x \notin S_i$ we have $u_{eq}(t) = u(t, x(t))$. So, the only question is, how to define $u_{eq}(t)$ on a switching surface. The scheme presented in [1] is based on resolving the system of algebraic equations:

$$\nabla^T s_i(x) f(t, x, u_{eq}) = 0 \text{ if } x \in S_i,$$

$$(u_{eq})_i = u_i(t, x) \qquad \text{if } x \notin S_i,$$

where $i = 1, 2, \ldots, m$ and $u_{eq} \in \mathbb{R}^m$ is a vector to be defined. If the obtained solution $u_{eq}(t, x)$ belongs to $U(t, x)$, then it is called *equivalent control* [1].

The next lemma reduces the existence analysis of the Utkin solution to analysis of a differential inclusion.

Lemma 1 ([14], p. 78) *Let a function $f : R^{n+m+1} \to R^n$ be continuous and a set-valued function $U : R^{n+1} \to 2^{\mathbb{R}^m}$ be defined and upper semicontinuous on an open set $I \times \Omega$, where $\Omega \subseteq R^n$. Let $U(t, x)$ be non-empty, compact, and convex for every $(t, x) \in I \times \Omega$. Let a function $x : R \to R^n$ be absolutely continuous on I, $x(t) \in \Omega$ for $t \in I$ and*

$$\dot{x}(t) \in f(t, x(t), U(t, x(t))), \tag{2.9}$$

almost everywhere on $\mathcal{I}$. Then, there exists a measurable function $u_{eq} : R \to R^m$ such that $u_{eq}(t) \in U(t, x(t))$ and

$$\dot{x}(t) = f(t, x(t), u_{eq}(t))$$

almost everywhere on I.

In general, the differential inclusion (2.9) may have non-convex right-hand side. Unfortunately, the existence of solutions for non-convex differential inclusions is a non-trivial problem [15]. Therefore, the algebraic approach introduced by equivalent control method is still the most effective way to find the Utkin solutions.

2.5 Sliding Mode Equations in Control Affine Systems

It is interesting that the both methods lead to the same sliding mode equations for affine (linear with respect to control) systems

$$\left. \begin{aligned} \dot{x}(t) &= f(t, x(t)) + B(t, x(t)) u(t, x(t)), \quad x(t) \in R^n \\ u_i(t, x) &= \begin{cases} u_i^+(t, x) & \text{if } s_i(x) > 0 \\ u_i^-(t, x) & \text{if } s_i(x) < 0 \end{cases} \\ u &= (u_1, \ldots, u_m)^\mathsf{T}, \quad s = (s_1, \ldots, s_m)^\mathsf{T} \end{aligned} \right\} \tag{2.10}$$

where $f : R^{n+k+1} \rightarrow R^n$ is a continuous vector-valued function, $B : R^{n+1} \rightarrow R^{n \times m}$ is a continuous matrix-valued function, and $u : R^{n+1} \rightarrow R^m$ is a piecewise continuous function, such that u_i has a unique time-invariant switching surface $s_i(x) = 0$, where $s_i : R^n \rightarrow R$ is a smooth function.

Theorem 1 *Definitions of Filippov and Utkin are equivalent if and only if*

$$\det\left(\nabla^T s(x) B(t, x)\right) \neq 0 \quad for \quad (t, x) \in S, \tag{2.11}$$

where

$$s(x) = (s_1(x), s_2(x), \ldots, s_m(x))^T, \quad \nabla s(x) \in R^{n \times m}$$

$s(x) = (s_1(x), s_2(x), \ldots, s_m(x))^T, \nabla s(x) \in R^{n \times m}$ *is the matrix of partial derivatives* $\dfrac{\partial s_j}{\partial x_i}$, *and S is a discontinuity set of $u(t, x)$.*

If the switching surface satisfies the condition of this theorem, then the equivalent control and sliding motion equation can be constructed analytically. For instance, the sliding motion condition $s = 0, \dot{s} = 0$ implies

$$\nabla^T s(x) f(t, x) + \nabla^T s(x) B(t, x) u(t, x) = 0. \tag{2.12}$$

The condition (2.12) allows us to obtain the equivalent control

$$u_{eq}(t, x) = -\left[\nabla^T s(x) B(t, x)\right]^{-1} \nabla^T s(x) f(t, x) \tag{2.13}$$

as well as the sliding mode equation

$$\dot{x}(t) = f_0(t, x(t)), \quad f_0 = \left(I_n + B\left[\nabla^T s B\right]^{-1} \nabla^T s\right) f. \tag{2.14}$$

According to Filippov's method, right-hand side f_0 of the sliding mode equation should belong to the convex hull of all possible vectors of the original system (2.10). Each control component can take two values; hence, the total number of all possible control vectors is equal to 2^m, $\{v^1, \ldots, v^{2^m}\}$. The intersection of convex hull of the right-hand side of (2.10)

$$\left\{ f(t, x) + B(t, x) z : z = \sum_{i=1}^{2^m} \mu_i v^i, 0 \leq \mu_i \leq 1, \sum_{i=1}^{2^m} \mu_i = 1 \right\}$$

with the tangent manifold $s(x) = 0$ defines the sliding mode equation. It follows that the equation

$$\nabla^T s(x) f(t, x) + \nabla^T s(x) B(t, x) z = 0$$

holds. Substitution of its solution

$$z = -\left[\nabla^T s(x) B(t, x)\right]^{-1} \nabla^T s(x) f(t, x)$$

into the equation for the convex hall results in the sliding mode equation. It coincides with Eq. (2.14) resulting from the equivalent control method.

Briefly describe the approaches to derive SM equation, different from Filippov and equivalent control methods.

Input $s(x)$ is equal to zero in sliding mode for the element implementing discontinuous control, while its output is different from zero. Based on this fact [16], offered to replace it by a linear element with gain tending to infinity and take the slow motion equation as a model of sliding mode. In [17], there was offered to find the solution with respect to s in convolution form for LTI system and then the control from the integral equation resulting from $s(t) = 0$. The sliding mode equation is obtained after the substitution of this control into the original system. The authors of [18] assumed that control in (2.7) takes all values between u^+ and u^-, while the authors of [19] offered to select the values of control from some set, which, on one hand, may not include all values between u^+ and u^- but, on the other hand, may be beyond this interval. In both cases, the authors use differential inclusion form $\dot{x} \in F(x, t)$, where $F(x, t)$ is a convex set thus introduced the right-hand side of the original system.

Although the above methods may lead to different sliding mode equations, the question "which one is correct?" cannot be asked, since all of them were postulated.

Variety of methodologies to analyze sliding mode equations can be found in excellent survey papers [20, 21].

2.6 Regularization

As demonstrated in previous sections, different ways of solution continuation may lead to different sliding mode equations except for non-singular affine systems. It is of interest to answer the natural question, which motion equation is implemented in a real-life system. Formally, the question cannot be answered. Systems with discontinuous control do not satisfy the conventional existence-uniqueness theorems since the Lipschitz constant does not exist for discontinuous systems. Any solution ambiguity may be excluded, if the system behavior is analyzed based on a more accurate model. The accurate model takes into account the way of implementation of discontinuous control such as hysteresis, time-delay, and high-gain amplifier with output limitation. Then the solution of the corrected ideal model exists and unique. If the limit of the solution exists with a small parameter, added to the ideal model and tending to zero, then it can be taken as the solution to the ideal sliding mode equations. This method is called *Regularization* [1, Chap. 2].

If different ways of regularization lead to different motion equations, we should admit that the equations beyond discontinuity surfaces do not allow us to derive the sliding mode equations unambiguously.

An interesting example of the relay system

$$\begin{aligned}
\dot{x}_1 &= 0.3x_2 + x_1 u, & s &= x_1 + x_2, \\
\dot{x}_2 &= -0.7x_1 + 4x_1 u, & u &= -\text{sign}(x_1 s)
\end{aligned} \right\} \tag{2.15}$$

with different sliding mode equations depending on the way of regularization is given in [1]. The solution exists for this system, if the ideal relay is replaced by a relay with hysteresis or by a linear function in small vicinity of sliding surface (Figs. 2.11 and 2.12). This example corresponds to Fig. 2.10 and explains why it is happened. The regularization method implies that the width of hysteresis loop or the vicinity with the linear function tends to zero. The limit procedure results in Filippov's equation with unstable solution in the first case and in equivalent control method equation with stable solution in the second case (Fig. 2.13). Further, the method by Filippov will be modified slightly to get all possible solutions depending on the way of regularization.

Fig. 2.11 Relay with hysteresis

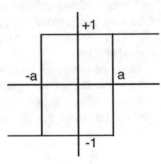

Fig. 2.12 Relay with limiter

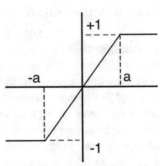

Fig. 2.13 Dependence of stability of regularization method: black line—solution before reaching of the sliding surface; green line—solution derived by means of equivalent control method; red line—Filippov solution

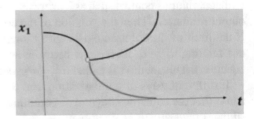

Filippov's method was offered as a postulate. And it represents the following rather evident fact. If the velocity vector can take several values in a vicinity of some point in the state space, then due to switching, the right-hand side of the motion dynamics is their minimal convex hull, and any other motions cannot appear. Generally speaking, depending on a way of the regularization, control u can take additional values (not u^+ or u^- only) and all possible sliding mode equations are defined by the convex hull, depending on all these values.

Regularization implies that the ideal control is replaced by a new one u_ε,

$$u_\varepsilon(t, x) = \begin{cases} u(t, x) & \text{if } x \notin N + \varepsilon B, \\ u_{reg}(t, x, \varepsilon) & \text{if } x \in N + \varepsilon B, \end{cases} \quad \varepsilon > 0, \quad (2.16)$$

where $N = \{x \in R^n : s(x) = 0\}$ is the discontinuity set of measure zero, B is the unit ball, and u_{reg} depends on a regularization method and selected such that some conventional uniqueness-existence theorem can be applied. Then, the set of all possible sliding mode Eq. (2.6), taking into account the regularization, can be found in terms of u_ε

$$F(t, x) = \lim_{\varepsilon \to 0} \text{Conv}\{f(t, x, u_\varepsilon(t, \{x + \varepsilon B\} \backslash N))\}.$$

For continuous approximation of a discontinuous function, the convex hull should be found for all possible values of u_{reg} (dashed sector in Fig. 2.14). Indeed, consider the set of sliding mode equations, resulting from the regularization approach, for the system (2.15). Suppose that

$$u_\varepsilon = \begin{cases} u^- & \text{if } s < 0, \\ u^+ & \text{if } s > 0. \end{cases}$$

As it can be seen in Fig. 2.14, sliding mode exists on $s = 0$. If the switching device is implemented with hysteresis, this regularization leads to Filippov method [1]. The intersections of the straight line connecting $f(x, t, u')$ and $f(x, t, u'')$ with the tangential line determine the sliding mode equation. It is evident that any point on the tangential line between $f_{filippov}$ and f_{eq} can be obtained by a proper choice of u^- and u^+. It means that the set of all possible velocity vectors $f_0 = \text{Conv}_u f(x, t, u)$ can be found in the intersection of the convex hull of the arc between u^- and u^+ (dashed sector in Fig. 2.14) and tangential line. Note that the sliding mode equation is not postulated but follows from the regularization method. Intersection of the hull with the manifold, tangential to the sliding manifold, determines all possible sliding mode vector fields, namely, all vector between $f_{filippov}$ and $f(x, u_{eq})$ in Fig. 2.14. Note that for this way of regularization they can be obtained from definition given in [18], assuming that $u^- \leq u_{reg} \leq u^+$. But it is not the case always, for example, for Coulomb friction [19].

Finally, it would be of interest to derive a class of systems with unique sliding mode equation for any way of regularization, or for any function u_{reg}. The method (2.16) is called "*boundary layer regularization*". It was formulated in [1, Chap. 2] in a slightly

Fig. 2.14 Regularization method

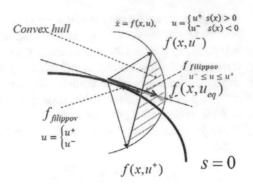

different way. The solution of the original system with control (2.16) depends on u_{reg} and ε. If for any function u_{reg} the following limit $\lim_{\varepsilon \to 0} x(t, u_{reg}, \varepsilon) = x(t)$ exists and unique, then this function $x(t)$ is taken as a solution to the motion equation in sliding mode. As shown above, generally speaking it is not the case. Filippov method and *equivalent control method* result in the same sliding mode equations (except for singular cases) for affine systems. The validity of those equations was substantiated for affine systems via the boundary layer regularization method [1]. It means that the sliding mode is governed by (2.8) for any control implementation way. Note that the system order can be reduced to $n - m$. Indeed, m state variables can be found as functions of the remaining $n - m$ ones from the algebraic equation $s(x) = 0$ (as follows from [1] the solution exists). Then, m variables are substituted into sliding mode equations, and finally the equations with respect to m variables should be excluded from the system.

References

1. Utkin, V.: Sliding Modes in Control and Optimization. Springer, Berlin (1992)
2. Utkin, V., Guldner, J.: Sliding Mode Control in Electro-Mechanical Systems. CRC Press (2009)
3. Edwards, C., Spurgeon, S.: Sliding Mode Control: Theory and Applications. Taylor and Francis (1998)
4. Shtessel, Y., Edwards, C., Fridman, L., Levant, A.: Sliding Mode Control and Observation. Control Engineering Series, Birkhauser, NY (2014)
5. Drazenovic, B.: The invariance conditions in variable structure systems. Automatica **5**(3), 287–295 (1969)
6. Glad, S.T.: Output dead-beat control for nonlinear systems with one zero at infinity. Syst. Control Lett. **9**, 249–255 (1987)
7. Utkin, V.I.: Variable structure and lyapunov control. In: Sliding Mode Control in Discrete-Time and Difference Systems. Springer, Berlin (1993)
8. Barbashin, E.A.: Introduction to Stability Theory. Nauka, Moscow (1970) (in Russian)
9. Blanchard, P., Devaney, R.L., Hall, G.R.: Differential Equations. Thompson Brooks/Cole (2006)
10. Levant, A.: Sliding order and sliding accuracy in sliding mode control. Int. J. Control **58**(6), 1247–1263 (1993)

11. Filippov, A.F.: Differential Equations with Discontinuous Righthand Sides. Kluwer Academic Publishers (1988)
12. Orlov, Y.: Discontinuous Systems: Lyapunov Analysis and Robust Synthesis Under Uncertainty Conditions. Springer, Berlin (2008)
13. Polyakov, A., Fridman, L.: Stability notions and lyapunov functions for sliding mode control systems. J. Frankin. Inst. **351**(4), 1831–1865 (2014)
14. Filippov, A.F.: On certain questions in the theory of optimal control. J. SIAM Control **1**(1), 76–84 (1962)
15. Bartolini, G., Zolezzi, T.: Variable structure systems nonlinear in the control law. IEEE Trans. Autom. Control **30**(7), 681–684 (1985)
16. Tsypkin, Ya.: Theory of Relay Automstic Regulations. State Publisher of Engineering Literature (1956)
17. Neimark, Y.I.: Note on A. Filippov Paper. In: Proceedings of 1st IFAC Congress II. Butterworths, London (1961)
18. Aizerman, M.A., Pyatnitskii, E.S.: Foundations of the theory of discontinuous systems. Autom. Remote. Control. **8**, 39–61 (1974)
19. Gelig, A.Kh., Leonov, G.A., Yakubovich, V.A.: Stability of Nonlinear Systems with Nonunique Equilibrium Position. Nauka (1978) (in Russian)
20. Cortes, J.: Discontinuous dynamic systems. IEEE Control. Syst. Mag. **28**(3), 36–73 (2008)
21. Leonov, G.A., Kuznetsov, N.V., Kiseleva, N.A., Mokaev, R.N.: Global problems for differential inclusions. Differ. Equ. **53**(13), 1671–1697 (2017)

Chapter 3
Design Principles

Abstract This chapter discusses the main principles of sliding mode design. Regular form of the controlled system is introduced. Reaching and existence conditions are presented. The decoupling procedure and invariance notion are discussed. The basic principles of output, unit, and integral SMC are presented. Second-order sliding mode (SOSM) controllers (such as twisting and super-twisting) are analyzed. Suboptimal controllers are also discussed.

Keywords Regular form · Reaching and existence conditions · Decoupling procedure · SOSM control · Twist and super-twist controllers

3.1 Sketch of the Design Procedure

The sliding mode equations based on an ideal model can be obtained unambiguously for affine systems. Therefore, the methodology for designing feedback sliding mode control was developed mainly for affine systems governed by Eq. (2.7). It is important to note that

- the sliding mode equation is of a reduced order;
- it does not depend on the control values;
- it depends on discontinuity manifold equation (2.8).

These properties allow us to outline the design procedure for the sliding mode control:

1. Select function $s : R^n \to R^m$ such that

$$|\det(\nabla^{\mathsf{T}} s \cdot B)| > \Delta_0$$

and the sliding mode on the manifold

$$s(x) = 0, x \in R^n$$

has desired properties. Any design method of the control theory (such as eigen-value placement and optimality) can be utilized for this reduced-order problem.

Remark 3.1 Step 1 implies that matrix $\nabla^\mathsf{T} s \cdot B$ is non-singular. It means that the relative degree between control as an input and vector s as an output is equal to 1. If it is not the case, the control is selected as a function of s and its time derivatives to make the system relative degree between control as an input and vector s as an output equal to 1. It was done at the early stage of SMC, when the system behavior was studied in the space of the output and its time derivatives. Later, the same method was used in the publications on high-order sliding mode control [1].

Remark 3.2 "Select s such that the sliding mode has desired properties". It is not the conventional problem statement, since the right-hand side depends on not only s, but also the gradients of its components.

Remark 3.3 Formally, sliding mode equations are equations of zero dynamics of the system with control and s as input and output.

2. Discontinuous control is selected to enforce sliding mode in the preselected manifold $s(x) = 0$. Again, we deal with the problem of a reduced order, since the state trajectories should reach the manifold after a finite time interval in m-dimensional subspace s in compliance with the equation

$$\dot{s} = \nabla^\mathsf{T} s \cdot f + \nabla^\mathsf{T} s \cdot Bu. \tag{3.1}$$

The above two-step design methodology with sliding mode in $(n-m)$-dimensional manifold was called "*Conventional sliding mode control*" in [2]. Further, this term will be used.

3.2 Regular Form

The design procedure becomes transparent, for the systems in the so-called "regular form" [3]

$$\left.\begin{array}{l} \dot{x}_1(t) = f_1(t, x_1(t), x_2(t)), \\ \dot{x}_2(t) = f_2(t, x_1(t), x_2(t) + B_2(t, x_1(t), x_2(t))u(t), \end{array}\right\} \tag{3.2}$$

where

$$x_1(t) \in R^{n-m}, \ x_2(t) \in R^m, \ u(t) \in R^m,$$
$$f_1 : R^{n+1} \to R^{n-m}, \ f_2 : R^{n+1} \to R^m,$$
$$B_2 : R^{n+1} \to R^{m \times m} \text{ and } \det(B_2) \neq 0.$$

The state component $x_2(t)$ is handled as a fictitious control input in the first equation of (3.2). Select

$$x_2 = -s_0(x_1)$$

to get the desired dynamics in the first subsystem. As a result, we deal with the design problem in the system of the $(n - m)$th order with m-dimensional control $x_2(t) = -s_0(x_1(t))$, where s_0 is a smooth function. Then, the discontinuous control should be designed to enforce sliding mode in the manifold

$$s(x_1, x_2) = x_2 + s_0(x_1) = 0, \quad x_1 \in R^{n-m}, \quad x_2 \in R^m$$

(the design problem of the mth order with m-dimensional control).

If sliding mode in the manifold $s(x_1, x_2) = 0$ starts after a finite time interval, then the system exhibits the desired behavior governed by

$$\dot{x}_1(t) = f_1(t, x_1(t), -s_0(x_1(t))).$$

Note that the motion is of a reduced order and depends neither on function f_2 nor on function B_2 in the second equation of the original system (3.2).

As mentioned in Remark 3.2, for a general case sliding mode equation depends on not only function s, but also the gradients of its components. The first step of the design procedure for systems in the regular form does not depend on gradients and the conventional methods of the control theory can be applied directly. The necessary and sufficient conditions for an affine system to be reduced to the regular form (3.2) can be found in [3] and [4, Chap. 6].

To demonstrate the sliding mode control design methodology, consider the conventional problem of linear control theory: *eigenvalue placement* in a linear time-invariant system:

$$\dot{x}(t) = Ax(t) + Bu(t),$$

where $x(t)$ and $u(t)$ are n- and m-dimensional state and control vectors, respectively, A and B are constant matrices, rank$(B) = m$. The system is assumed to be controllable. For any controllable system, there exists a linear feedback $u = Fx$ (F being a constant matrix), such that the eigenvalues of the feedback system, i.e., of matrix $A + BF$, take the desired values: as a result, the system exhibits desired dynamic properties. Reducing system equations to the regular form will be performed as a preliminary step in the sliding mode control design procedures. Since rank $= m$ matrix B may be partitioned (after reordering the state vector components) as

$$B = \begin{bmatrix} B_1 \\ B_2 \end{bmatrix},$$

where

$$B_1 \in R^{n-m \times m}, \quad B_2 \in R^{m \times m}, \quad \det(B_2) \neq 0.$$

The non-singular coordinate transformation

$$\begin{bmatrix} x_1 \\ x_2 \end{bmatrix} = Tx, \quad T = \begin{bmatrix} I_{n-m} & -B_1 B_2^{-1} \\ 0 & B_2^{-1} \end{bmatrix}$$

reduces the system equations to the regular form

$$\dot{x}_1(t) = A_{11}x_1(t) + A_{12}x_2(t),$$
$$\dot{x}_2(t) = A_{21}x_1(t) + A_{22}x_2(t) + u(t),$$

where $x_1(t) \in R^{n-m}$, $x_2(t) \in R^m$, and A_{ij} are constant matrices, $i, j = 1, 2$. It follows from the controllability of (A, B) that the pair (A_{12}, A_{22}) is controllable as well [4]. Handling x_2 as an m-dimensional intermediate control in the controllable $(n - m)$-dimensional first subsystem, all eigenvalues may be assigned arbitrarily by a proper choice of matrix $C \in R^{m \times (n-m)}$ in

$$x_2 = -Cx_1.$$

To provide the desired dependence between components x_2 and x_1 of the state vector, sliding mode should be enforced in the manifold

$$s(x_1, x_2) = x_2 + Cx_1 = 0,$$

where $s = (s_1, s_2, \ldots, s_m)^\mathsf{T}$ is the difference between the real values of x_2 and its desired value $-Cx_1$. After sliding mode starts, the motion is governed by the system with the desired eigenvalues

$$\dot{x}_1(t) = (A_{12} - A_{12}C)x_1(t).$$

Now, eigenvalue problem can be solved easily for the system of a reduced order, since the selection of matrix C is a conventional design problem of the linear control theory.

3.3 Reaching and Existence Conditions

Although the design procedure in Sect. 3.1 consists of two problems of reduced orders, it cannot be stated that the design problem is decoupled into two independent subproblems. Indeed, Eq. (3.1) depends on function s selected at the first step as well. The second step is selection of discontinuous control such that the sliding mode is enforced in manifold $s(x) = 0$. It means that for any initial conditions the state should reach $s(x) = 0$ and it should be a sliding manifold according to Definition 2.1.

The reaching and sliding mode existence conditions can be derived based on Eq. (3.1). Let the ith component of control undergoes discontinuities on surface $s_i(x) = 0$

$$u_i(t, x) = \begin{cases} u_i^+(t, x) & \text{if } s_i(x) > 0, \\ u_i^-(t, x) & \text{if } s_i(x) < 0, \end{cases} \quad u_i^+(t, x) \neq u_i^-(t, x),$$

for example,

$$u_i(t, x) = u_i^0(t, x) + u_i^1(t, x)\text{sign}(s_i(x)),$$
$$u_i^0 = 0.5(u_i^+ + u_i^-),\ u_i^1 = 0.5(u_i^+ - u_i^-)$$

and

$$\left.\begin{array}{c} \dot{s}(t) = d(t, x(t)) - D(t, x(t))\text{sign}(s(x(t))), \\ d = \nabla^\mathsf{T} s\ (f + Bu^0),\quad D = -\nabla^\mathsf{T} s\ B\text{diag}\{u_i^1\}, \\ u^0 = (u_1^0, u_2^0, \ldots, u_m^0)^\mathsf{T}. \end{array}\right\}$$
(3.3)

Theorem 3.1 ([4]) *If the matrix $D + D^\mathsf{T} > 0$ is positive definite, then there exists scalar function $M_0 : R^{n+1} \to R$, $M_0 > 0$ such that manifold $s(x) = 0$ is reached after a finite time interval and $s(x) = 0$ is a sliding manifold for control*

$$u = u^0 + M\,\text{sign}(s)$$

with $M : R^{n+1} \to R$ and $M > M_0$.

As mentioned above, the two steps of the control design are interconnected since matrix ∇s depends on the preselected function s. Generally speaking, $D + D^\mathsf{T}$ is not positive definite.

3.4 Decoupling

Manifold $s(x) = 0$ is the intersection of m surfaces $s_i(x) = 0$ selected such that the sliding mode has the desired properties. Manifold $s(x) = 0$ and as a result the sliding mode equations are the same if vector $s(x)$ is replaced by

$$s^*(t, x) = Q(t, x)s(x),\ \det(Q) \neq 0,$$

and the control undergoes discontinuities on the surface $s^*(t, x) = 0$. However, the motion in subspace s depends on matrix Q.

The idea of decoupling is in finding matrix $Q(t, x)$ for any preselected vector $s(x)$ such that the state trajectories are converging to the origin in subspace s. Equation (3.1) can be written in the form

$$\dot{s} = [\nabla^\mathsf{T} s \cdot B](u - u_{eq}),$$

where u_{eq} was found for the system (2.7). Calculate time derivative of Lyapunov function $V = 0.5s^\mathsf{T} s$ along the trajectories of the system (3.1):

$$\dot{V} = s^\mathsf{T}[\nabla^\mathsf{T} s \cdot B](u - u_{eq}).$$

Then, replace the switching manifold by a new one

$$s^*(t, x) = 0, \ s^* = [\nabla^\mathsf{T} s \cdot B]^\mathsf{T} s.$$

For control $u = -M \, \mathrm{sign}(s^*)$, $M > 0$, we have

$$\dot{V} = -M|s^*| - (s^*)^\mathsf{T} u_{eq} \leq -|s^*|(M - |u_{eq}|).$$

Under a natural assumption that the upper estimate $u_{\max} > |u_{eq}|$ for the equivalent control is known, $\dot{V} < 0$ and all trajectories are converging to manifold $s(x) = 0$ (or to $s^*(t, x)$), if $M > u_{\max}$. Finally, it can be shown that the state reaches $s(x) = 0$ after a finite time interval provided that

$$\dot{V} \leq -\alpha \|s^*\|, \quad \alpha = \min_{(t,x)}[M(t, x) - u_{\max}(t, x)].$$

Indeed, since

$$\|s^*\| = \|[\nabla^\mathsf{T} s \cdot B]^\mathsf{T} s\| \geq \sqrt{\lambda_{\min}} \|s\|,$$

where $\lambda_{\min} : R^{n+1} \to R$ is a minimal eigenvalue of the positive symmetric matrix

$$\Xi = [\nabla^\mathsf{T} s \cdot b] \cdot [\nabla^\mathsf{T} s \cdot B]^\mathsf{T} \ : \ \det(\Xi) \geq \xi > 0,$$

then

$$\dot{V} \leq -\gamma\sqrt{V}, \quad \gamma = \alpha \sqrt{2 \inf_{(t,x)} \lambda_{\min}(t, x)},$$

and the sliding manifold $s(x) = 0$ is reached in a finite time

$$t_{reach} \leq \frac{\sqrt{2} \, \|s(x(0))\|}{\gamma}$$

(see, for example, [5, Chap. 2]).

3.5 Invariance

Assume that for an arbitrary affine system

$$\dot{x}(t) = f(t, x(t)) + B(t, x(t))u(t) + h(t, x), \tag{3.4}$$

the function h characterizes the disturbances and parameter variations which should not affect the feedback system dynamics. In compliance with the equivalent control method, the solution to

$$\dot{s} = \nabla^\mathsf{T} s \, (f + Bu + h) = 0$$

with respect to control,

$$u_{eq} = -[\nabla s^\mathsf{T} \cdot B]^{-1} \nabla s (f + h)$$

should be substituted into (3.4) to yield the sliding mode equation

$$\dot{x}(t) = f_0(t, x(t)), \quad f_0 = \left(I_n - B \left[\nabla^\mathsf{T} s \, B \right]^{-1} \nabla^\mathsf{T} s \right)(f + h). \qquad (3.5)$$

Let range of $B(t, x)$ be a subspace formed by the base vectors of matrix $B(t, x)$ for each point (t, x). The sliding mode is invariant with respect to function h if

$$h(t, x) \in \text{range}(B(t, x)), (t, x) \in R^{n+1}.$$

This condition means that there exists function γ such that $h = B\gamma$. Direct substitution of function h into (3.5) demonstrates that the sliding motion in any manifold $s(x) = 0$ does not depend on perturbation function h. As it follows from the design methods of this section, an upper estimate of this vector is needed to enforce the sliding motion. Condition

$$h(t, x) \in \text{range}(B(t, x))$$

is called *matching condition,* and it is a natural generalization of the invariance condition obtained in [6] for linear systems.

The invariant SMC can be designed, if a state vector is available; however, it is not the case always, and an output can be measured only. The traditional approach implies designing a state observer, but in our case it should be done under uncertainty conditions (vector $h(t, x)$ is unknown). The solution was offered in the set of publications by H. Khalil for the special case, when motion equations are in canonical space, consisting of an output and its time derivative or a set of similar subspaces, if an output is a vector [7, 8]. All state variables can be found by the observer with the appropriate hierarchy of its input gains tending to infinity. It is important that the offered method does need the unknown disturbance. The stability of the overall system consisting of the observer and the plant is analyzed in [8].

3.6 Output Sliding Mode Control

The design methods of the previous section has been discussed under the assumption that all state components are available. In most practical situations, an output vector can be measured only. The objective of this section is to derive the class of systems, which can be stabilized by sliding mode control utilizing a system output only. The design methodology for a special case of systems with the same dimension of control and output [9] will be considered. The objective of this subsection is to single out the class of systems with matched uncertainties

$$\dot{x}(t) = Ax(t) + B[k(t,x)u(t) + h(t,x)],$$
$$y(t) = Cx(t),$$
$$x(t) \in R^n, \ u(t) \in R^m, \ y(t) \in R^m, \ \mathrm{rank}(B) = \mathrm{rank}(C) = m,$$
$$0 < k_0 \le k(t,x) \in R, \ \|h(t,x)\| \le h_0,$$

assuming, that

- dimensions of the output and the control input are the same $y(t) \in R^m$, $u(t) \in R^m$,
- $\det(CB) \ne 0$,
- sliding mode is enforced in manifold $y = 0$ and sliding motion in this manifold has some desired properties such as stability.

In contrast to the conventional two-step design procedure, the first step (selection of sliding manifold) is not needed, and the second step (enforcing sliding mode) should be solved using output y only. It is assumed that there exists a linear control stabilizing the system with no disturbances. Finally, the real control consists of this linear component and a discontinuous part to enforce sliding mode at the presence of the bounded disturbance $h(t,x)$, satisfying matching condition.

Now it is demonstrated, how this design idea is implemented. As in the previous subsection, the system can be reduced to the regular form

$$\dot{x}_1(t) = A_{11}x_1(t) + A_{12}x_2(t),$$
$$\dot{x}_2(t) = A_{21}x_1(t) + A_{22}x_2(t) + k(t,x)u(t) + h(t,x)),$$
$$y(t) = C_1 x_1(t) + C_2 x_2(t), \ x_1(t) \in R^{n-m}, \ x_2(t) \in R^m.$$

Condition $\det(CB) \ne 0$ means $\det(C_2) \ne 0$. Without loss of generality, sliding manifold can be written as $y = 0$ with

$$y = C_1 x_1 + x_2.$$

As in previous subsections, the system can be reduced to the regular form

$$\dot{x}_1(t) = (A_{11} - A_{12}C_1)x_1(t) + A_{12}y(t)),$$
$$\dot{y}(t) = (C_1(A_{11} - A_{12}C_1) + A_{21} - A_{22}C_1)x_1(t)$$
$$+(C_1 A_{12} + A_{22})y(t) + k(t,x)u(t) + h(t,x(t)).$$

Of course sliding mode equation

$$\dot{x}_1(t) = (A_{11} - A_{12}C)x_1(t)$$

in $y = 0$ is stable with a Lyapunov function defined as $x_1^{\mathsf{T}} P_1 x_1$, where P_1 is a positive-definite matrix. As shown in [9], using the Lyapunov function

$$V = x_1^{\mathsf{T}} P x_1 + y^{\mathsf{T}} y,$$

the system with control

$$u = -qy - M \operatorname{sign}(y), \; M > h_0/k_0$$

and high enough q is asymptotically stable. It follows from Sylvester criterion directly (see [10]). The final stage of the process is in sliding mode, since condition (2.2) for this motion to exist holds in the vicinity of the origin.

3.7 Integral SMC

3.7.1 Main Idea

The design procedure of SMC implies that the motion consists of two phases: reaching the preselected manifold in the state space and sliding along the manifold with the desired properties. *Integral sliding mode control* (ISMC) excludes the reaching phase and enforces the desired motion from an initial time instant [11], [5, Chap. 7].

Its design methodology can be demonstrated for the simple second-order system

$$\left. \begin{array}{l} \dot{x}_1(t) = x_2(t), \; x_1(t) \in R, \\ \dot{x}_2(t) = u + h(t, x), \; x_2(t) \in R, \end{array} \right\}$$

where the upper limit of disturbance $|h(t, x)| \le h_0$ is known only. First, the control is selected such that the system with $h = 0$ would be governed by the linear equation with roots of characteristic equation equal to $\{-1, -1\}$. It happens if control is equal to

$$u_0 = -x_1 - 2x_2.$$

Complement the original system by the third equation

$$\dot{z}(t) = -u_0(t), \quad z(0) = 0,$$

and select the control

$$u(t) = u_0(t) - M \operatorname{sign}(s(t)), \quad M > h_0.$$
$$s(t) = x_2(t) - x_2(0) + z(t).$$

Since

$$\dot{s}(t) = -M \operatorname{sign}(s(t)) + h(t, x)$$

and $s(0) = 0$, the sliding mode occurs at the initial instant of time and $(M \operatorname{sign}(s))_{eq} = h(t, x)$. Substitution of

$$u = u_0 - (M \operatorname{sign}(s))_{eq}$$

into the original system results in desired motion equation

$$\dot{x}_1(t) = x_2(t), \dot{x}_2(t) = -x_1(t) - 2x_2(t).$$

An affine system with matched uncertainties is considered

$$\begin{aligned} \dot{x}(t) &= f(t, x(t)) + B(t, x(t))u(t) + h(t, x), \ t > 0, \\ x(t) &\in R^n, \quad u(t) \in R^m, \quad h = B\gamma, \quad \text{rank}(B) = m. \end{aligned} \right\} \tag{3.6}$$

Bounded function h (or γ) represents all uncertainties. Only upper estimate $\gamma_0 > \|\gamma\|$ is known. It is assumed that control u_0 can be found for the nominal system (system (3.6) with $h = 0$) in correspondence with some criterion. Let $x_0(t), t > 0$ be a solution to the nominal feedback system

$$\dot{x}_0(t) = f(t, x_0(t)) + B(t, x_0(t))u_0(t, x_0(t)). \tag{3.7}$$

Control u should be designed such that the solution to (3.6) coincides with the solution to (3.7) starting from the initial time.

3.7.2   Design Method

Select the control in the form

$$\begin{aligned} u(t) &= u_0(t, x(t)) - M(t, x)(G(x(t)) B(t, x(t)))^{-1}\text{sign}(s(t)), \\ s(t) &= s_0(x(t)) - s_0(x(0)) + z(t), \\ \dot{z}(t) &= -[G(x(t))f(t, x(t)) + G(x(t))B(t, x(t))u_0(t, x(t))], \\ G(x) &= -\frac{\partial s_0}{\partial x}, \quad z(0) = 0. \end{aligned}$$

At initial time, the state is in the manifold $s = 0$ in space (x, z), and function M will be selected such that sliding mode is enforced in the manifold. Let $s_0(x)$ be an arbitrary function satisfying condition

$$\det[G(x)B(t, x)] \neq 0, t \geq 0, x \in R^n.$$

Remark 3.4 In our second-order example,

$$GB = 1, \ f = 0, \ s_0 = x_2.$$

As it follows from

$$\dot{s} = GB(u + \gamma) = -M\text{sign}(s) + GB\gamma,$$

sliding mode in the manifold $s = 0$ exists if $M > \|GB\|\gamma_0$. The equivalent control

$$u_{eq}(t, x(t)) = -\gamma(t, x(t))$$

is the solution to $\dot{s}(t) = 0$. Substitution of u_{eq} into (3.6) results in sliding mode equation

$$\dot{x}(t) = f(t, x(t)) + Bu_0(t, x(t)).$$

It coincides with the desired equation for the system without uncertainties (3.7). The desired motion occurs from initial time. Note that in contrast to the traditional design methodology, the order of this motion is not reduced.

As an example, consider linear system

$$\dot{x}(t) = Ax(t) + Bu(t) + B\gamma(t, x(t)),$$

A, B are constant matrices, $\text{rank}(B) = m$, $\gamma_0 > \|\gamma\|$ and select control such that the system is governed by homogeneous time-invariant equation with desired eigenvalues starting from the initial time. If the system is controllable, then there exists matrix F such that matrix $A + BF$ has desired eigenvalues. Select control in the form

$$u(t, x(t)) = u_0(x(t)) + u_i(t, x(t)), \quad u_0(x) = Fx,$$
$$u_i(t, x) = -M \, \text{sign}(s_i(t, x)),$$
$$s(t, x(t)) = s_0(x(t)) - s_0(x(0)) + z(t), \quad s_0(x) = B^+x, \quad B^+B = I,$$
$$\dot{z}(t) = -B^+Ax(t) - u_0(x(t)).$$

Then

$$\dot{s} = -M \, \text{sign}(s) + \gamma$$

and sliding mode exists, if $M > \gamma_0$. Equivalent control

$$u_{i_{eq}}(t, x(t)) = -\gamma(t, x(t))$$

and sliding mode is governed by full-order equation starting from the initial time

$$\dot{x}(t) = (A + BF)x(t)$$

with the desired eigenvalues.

The application of ISM to LQ problem and specific algebraic observers may be found in [12].

3.8 Unit Control

The unit feedback synthesis has been developed to simplify the stability analysis and avoid extra computations required to bring the system into the regular form. The controller is synthesized in such a manner that the time derivative of a Lyapunov function, selected for a nominal, asymptotically stable system, remains negative definite in spite of parameter variations and external disturbances, which affect the system. The approach was developed in [13, 14], then was applied for designing sliding mode control, and was referred to as a *unit control* [15]. Its norm is equal to one everywhere except for the discontinuity manifold.

Similar to the component-wise control, the unit control with sufficiently high magnitude can enforce asymptotically stable sliding mode, which is robust against matched disturbances. The important point is that the trajectories of the closed-loop system never pass through the discontinuity manifold. The system stability is thus analyzed beyond the manifold. Once the trajectory is on the discontinuity manifold, smooth dynamics restore, and the standard Lyapunov theory is in force.

The unit control methodology was developed for affine systems

$$\dot{x}(t) = f(t, x(t)) + B(t, x(t))u(t) + h(t, x) \tag{3.8}$$

with matched disturbances such that $h = B\gamma$, $\|\gamma\| \le \gamma_0$.

The equation

$$\dot{x}(t) = f(t, x(t)) \tag{3.9}$$

represents an open-loop nominal system. For simplicity, the nominal system (3.9) is assumed to be asymptotically stable with some *a priori* known positive-definite continuously differentiable Lyapunov function V, such that its time derivative computed along the trajectories of (3.9) is negative definite,

$$W_0(t, x) = \nabla^{\mathsf{T}} V(x) f(t, x) \le -W_1(x), \quad (t, x) \in R^{n+1},$$

where W_1 is a continuous, positive-definite function. The time derivative of V on the trajectories of the perturbed system (3.8) is of the form

$$W = \frac{dV}{dt} = W_0 + \nabla^{\mathsf{T}} V \cdot B(u + \gamma).$$

Let the system (3.8) be driven by the control input

$$u = -\rho(t, x)U(s(t, x)), \quad U = \frac{s}{\|s\|} \tag{3.10}$$

with a scalar function ρ such that

$$\rho > \gamma_0, \tag{3.11}$$

and m-vector function

$$s(t, x) = B^{\mathsf{T}}(t, x)\nabla V(x).$$

Control (3.10) is called the unit control since $\|U\| = 1$ everywhere beyond $s = 0$ where it undergoes a discontinuity.

Then due to (3.11), the time derivative of the Lyapunov function $V(x)$, computed on the trajectories of the closed-loop system (3.8), is negative definite:

$$W = W_0 - \rho \parallel B^{\mathsf{T}} \nabla V \parallel + \nabla^{\mathsf{T}} V \cdot B\gamma \le$$
$$-W_1 - \parallel B^{\mathsf{T}} \nabla V \parallel [\rho - \gamma_0] \le -W_1.$$

It means that the closed-loop system is asymptotically stable.

It is of interest to note that in contrast to the conventional sliding mode control signals, which undergo discontinuities whenever a component of the sliding manifold changes sign, the unit control action is a continuous state function until the manifold $s = 0$ is reached. Due to this difference, the unit control method is an appropriate tool of discontinuous control design not only in a finite-dimensional state space, but also in an infinite-dimensional state space where control inputs are not (or even cannot be) represented in a component-wise form.

3.9 Second-Order Sliding Mode Control

The second-order sliding mode control (SOSM) was originated for systems with scalar input

$$\left. \begin{array}{l} \dot{x}(t) = f(t, x(t)) + b(t, x(t))u(t), \\ x(t) \in R^n, \quad u(t) \in R. \end{array} \right\} \tag{3.12}$$

The design method consists of two steps similar to the conventional one: selection of a sliding manifold and selection of a discontinuous control, enforcing a sliding mode. But in contrast to the conventional method, the dimension of a sliding manifold is equal to $n - 2$, but not $n - 1$. It explains the title of the method. As before, a sliding manifold is chosen such that the motion along it exhibits the desired properties and the reaching problem should be solved. Since the order of sliding mode equation is equal to $n - 2$, the reaching problem is a finite-time stability one of a second-order nonlinear system.

We start with the second-order system with evident stability conditions (see Example 3.8 in [16]). As seen later, they can be applied directly to SOSM. The mechanical *"mass-spring-damper"* system (Fig. 3.1) with no Coulomb friction and with no external forces is asymptotically stable because its energy dissipates and it may therefore serve as a Lyapunov function. The right-hand side of the motion equation

Fig. 3.1 Mass-spring-damper system

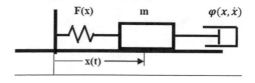

$$\ddot{x} = -\phi(x, \dot{x}) - F(x) \tag{3.13}$$

consists of viscous friction ϕ satisfying

$$\dot{x}\phi(x, \dot{x}) > 0$$

for $\dot{x} \neq 0$ and spring force F such that

$$xF(x) > 0$$

if $x \neq 0$. Lyapunov function is sum of kinetic and potential energies

$$V = \frac{m\dot{x}^2}{2} + \int_0^x F(y)dy.$$

The time derivative

$$\dot{V} = m\dot{x}\ddot{x} + F(x)\dot{x} = -\dot{x}\phi(x, \dot{x}) < 0$$

if $\dot{x} \neq 0$, which confirms our statement about stability. Note, V stops decaying only if $\dot{x} = 0$, but, as follows from (3.13), $\dot{x} \equiv 0$ means $x \equiv 0$ and asymptotic stability.

3.9.1 Twisting Algorithm

Illustrate the design idea for twisting algorithm [17] for the case $\nabla^\mathsf{T} s\, b = 1$

$$\dot{s} = \nabla^\mathsf{T} s\, f + \nabla^\mathsf{T} s\, b\, u. \tag{3.14}$$

Certainly, we deal with the conventional sliding mode control if u is a discontinuous function of s such that the sliding mode with the desired properties in $s = 0$ is enforced. The range of sliding mode control application will be increased, if the same effect will be obtained with continuous control actions, since not all actuators can operate with discontinuous inputs. For example, high-frequency switching may result in damage of valves in hydraulic actuators. A possible solution is evident: control signal is generated by the output of an auxiliary integrator with a discontinuous function of s and its time derivative as input. If this function is linear $cs + \dot{s}$, the origin in plane $(s, \dot{s})$ is reached asymptotically in the sliding mode. Again this design method is in a framework of the conventional sliding mode control.

Twisting algorithm

$$\left.\begin{array}{c} u(t) = v(t), \quad \dot{v}(t) = -M_1 \mathrm{sign}(s(t)) - M_2 \mathrm{sign}(\dot{s}(t)), \\ M_1, M_2 \text{ are positive constants,} \end{array}\right\} \tag{3.15}$$

Fig. 3.2 Part of the state
trajectory for twisting system

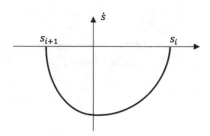

solves the same problem of reaching the origin in the plane $(s, \dot{s})$, but it is reached
after a finite time interval and there is no sliding mode in the reaching phase.

As follows from (3.14)

$$\ddot{s}(t) = d(t) - M_1 \text{sign}(s(t)) - M_2 \text{sign}(\dot{s}(t)), \quad d(t) = \frac{d}{dt} (\nabla^{\mathsf{T}} s \, f). \quad (3.16)$$

Parameters should satisfy inequalities

$$M_2 > |d|, \, M_1 > M_2 + |d|,$$

where d is a function of the states $s(x)$ and u, u is a state variable like x or s and
it is common for relay system to assume that the domain of initial conditions is
bounded (otherwise, unstable plant cannot be stabilized). Then $M_2 \text{sign}(\dot{s}) - d$ and
$M_1 \text{sign}(s)$ may be considered as function ϕ and F in Eq. (3.12) of the mechanical
system; therefore, the motion in space $(s, \dot{s})$ is asymptotically stable. Condition
$M_1 > M_2 + |d|$ means that sliding mode cannot appear on the line $\dot{s} = 0$, if $s \neq 0$,
because $\ddot{s}$ cannot change sign, which contradicts existence conditions (2.2).

And now about the finite-time convergence. Intuitively, the finite-time convergence can readily be explained from the mechanical standpoint. The rate of energy
dissipation is proportional to friction. It is always not less than $M_2 - |d|$, and there-
fore the energy is dissipated at a non-infinitesimal rate; and it will be equal zero after
final interval of time, and therefore both s and $\dot{s}$ will be equal to zero as well.

Finite-time convergence can be proven easily for the system with no disturbance.
The states cannot remain sign constant (Fig. 3.2). Finding the solution to Eq. (3.16)
with $d = 0$ and a piecewise constant right-hand side ($M_1 > M_2$) is a trivial problem.
The results of the calculations:

$$|s_{i+1}| = \gamma |s_i|,$$

where

$$s_i = s(t_i), \, \gamma = \frac{M_1 - M_2}{M_1 + M_2}, \, \Delta t_i = t_{i+1} - t_i = \delta \sqrt{|s_i|} = \delta \sqrt{\gamma^i |s_0|},$$

$$\delta = \sqrt{\frac{2}{M_1 - M_2} + \frac{\sqrt{2(M_1 - M_2)}}{M_1 + M_2}}.$$

Gain $\gamma < 1$, and therefore s converges to zero. Sequence Δt_i constitutes a geometric progression also with gain $\sqrt{\gamma} < 1$, which means finite-time convergence.

Finite-time convergence can be demonstrated based on the homogeneity condition. The right-hand side of the motion equation written in the form

$$\dot{y} = f(y), \quad y = \begin{pmatrix} y_1 \\ y_2 \end{pmatrix} = \begin{pmatrix} s \\ \dot{s} \end{pmatrix},$$

$$f(y) = \begin{pmatrix} y_2 \\ -M_1 \text{sign}(y_1) - M_2 \text{sign}(y_2) \end{pmatrix}$$

is a particular case of functions with the *homogeneity property*

$$f(Cy) = \beta C f(y), \quad C = \text{diag}\{c_i\}, i = 1, 2 \quad \beta, c_i \; - \; \text{scalar parameters.}$$

If the system is asymptotically stable and $0 < c_i < 1, \beta > 1$, the state converges to the origin in a finite time [18]. Function f satisfies the homogeneity condition with

$$\beta = 2, c_1 = 0.25, c_2 = 0.5.$$

Asymptotic stability was shown above based on mechanical analogy. It proves the finite-time convergence property of the twisting algorithm without disturbance. Note that asymptotic stability takes place in the system with bounded disturbances $d(y_1, y_2, t)$,

$$|d| \leq d_0 = \text{const},$$

if $M_2 > d_0$. Homogeneity conditions, complemented by $c_2\beta \geq 1$ (in our case $c_2\beta = 1$), mean finite-time convergence of the twisting algorithm with disturbances as well [19].

3.9.2 Super-Twisting Algorithm

As before, the control input of the super-twisting algorithm [17] (Fig. 3.3) is a continuous state function

$$\left. \begin{array}{l} u(t) = v(t) - a\sqrt{|s(t)|}\text{sign}(s(t)), \\ \dot{v}(t) = M \, \text{sign}(s(t)), \quad a, M \text{ are positive constants.} \end{array} \right\} \tag{3.17}$$

Formally, super-twisting control is two-dimensional, since real control and input of artificially introduced integrator

$$u_1 = -M \, \text{sign}(s)$$

Fig. 3.3 Super-twisting
control

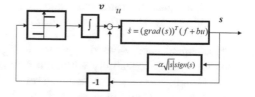

are selected by a designer. At the same time, super-twisting control is the only
example when time derivative of the output s is not needed for SOSM.

   Analyze stability of the motion in space $(s, \dot{s})$, governed by the second-order
equation

$$\ddot{s} = -a\frac{\dot{s}}{\sqrt{|s|}} - M\operatorname{sign}(s) + d. \tag{3.18}$$

Mechanical model can be utilized to prove asymptotic stability if $d = 0$. Indeed,
functions $a\dfrac{\dot{s}}{\sqrt{|s|}}$ and $M\operatorname{sign}(s)$ satisfy the conditions for ϕ and F in (3.13). As
shown in [17], the finite-time convergence takes place in the system in the presence
of bounded disturbances d.

   The finite-time convergence of systems with super-twisting control and no disturbances follows from the homogeneity property as well. Asymptotic stability again
was demonstrated based on the mechanical analogy in this subsection. The motion
equation can be written in the form

$$\dot{y} - f(y), \quad y = \begin{pmatrix} y_1 \\ y_2 \end{pmatrix} = \begin{pmatrix} s \\ \dot{s} \end{pmatrix},$$

$$f(y) = \begin{pmatrix} y_2 \\ -a\frac{y_2}{\sqrt{|y_1|}} - M\operatorname{sign}(y_1) \end{pmatrix}.$$

Again function f satisfies the homogeneity condition with

$$\beta = 2, \; c_1 = 0.25, \; c_2 = 0.5,$$

which proves finite-time convergence. The general form of homogeneity conditions
will be discussed in the next section along with application for super-twisting systems
with non-zero disturbances.

Remark 3.5 As in the conventional sliding mode, enforcing the second-order sliding
mode in the twisting algorithm implies that the relative degree between the discontinuous control and the input of the element, implementing this control, is equal to one,
since $\dot{s}$ is needed. The remarkable property of the super-twisting algorithm is that
the sliding mode is enforced with the relative degree between discontinuous input
$-M\operatorname{sign}(s)$ and the output s equal to two, and time derivative $\dot{s}$ is not needed. The
property takes place in a special case only, when the continuous part of the equation
does not satisfy the Lipschitz condition (function $\sqrt{|s|}$). Finally, as follows from

the motion equations, functions s and $\dot{s}$ cannot remain sign constant, and they have interlacing zeros. Therefore, the algorithms were called twisting and super-twisting.

Mechanical energy-based approach being applied to super-twisting Eq. (3.17) leads to asymptotic stability with zero disturbance only, since the both conditions $\dot{s}q_1(s, \dot{s}) > 0, sq_0(s) > 0$ are satisfied for $F(t) = 0$ as well. Equations (3.15), (3.17) with $F(t) = 0$ are particular cases of the above model of mechanical system with

$$q_0 = \bar{M} \operatorname{sign}(s), \quad \bar{M} = M_0 \text{ or } \bar{M} = M.$$

Lyapunov function for these particular cases in the form

$$V = \frac{1}{2}\dot{s}^2 + \bar{M}|s|$$

was offered in [20] or

$$V = \frac{1}{2}\dot{s}^2 + \bar{M}|s| + \frac{k}{2}s^2,$$

if control is complemented by a linear term ks.

3.10 Suboptimal Control

The so-called suboptimal controller [21, 22] can be classified as second-order sliding control, since the origin in plane $(s, \dot{s})$ is reached after a finite time interval the with no sliding modes at this interval. The design idea stems from time optimal control for system

$$\ddot{x} = u$$

with

$$|u| \leq M = \text{const.}$$

For initial conditions $x(0) > 0, \dot{x}(0) = 0$, the optimal process consists of two intervals. The state x decays from $x(0)$ to $x(0)/2$ at the first interval with $u = -M$ and then from $x(0)/2$ to 0 at the second interval with $u = M$. Apply the design methodology to the equation

$$\ddot{s}(t) = u(t) + d(t), \quad |d(t)| \leq d_0 = \text{const},$$
$$u(t) = \begin{cases} -M_1 & \text{if } s(t) - s^*/2 > 0, \\ M_2 & \text{if } s(t) - s^*/2 \leq 0, \end{cases} \qquad (3.19)$$

where $M_1 > 0, M_2 > 0$ are constant to properly be chosen according to [21], s^* is the value of s detected at the last time moment when $\dot{s}(t) = 0$. We define $s^* = 0$ at the initial instant of time. Note that the time instant when $\dot{s} = 0$ is needed for

Fig. 3.4 Suboptimal control

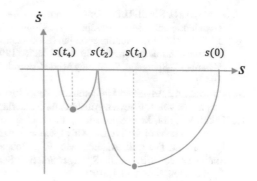

implementation (3.19) rather than time function $\dot{s}(t)$. It can be found estimating difference between $s(t)$ and $s(t + \delta)$ for small value of δ.

For initial conditions

$$s(0) > 0, \dot{s}(0) = 0$$

and $M_i > d_0$, the value of s is decaying until $s(t_1) = s(0)/2$ (Fig. 3.4). Control u is equal to M_2 for $t > t_1$ until $\dot{s}(t_2) = 0$. The gains M_1 and M_2 are properly chosen according to [22]. It is shown that for all t_{2i} and $\Delta t_i = t_{2i} - t_{2i-2}, i = 1, 2, \ldots$ the conditions

$$s(t_{2i}) < \gamma s(t_{2i-2}),$$
$$\delta \Delta t_i < \delta \Delta t_{i-1}, 0 < \gamma < 1 \text{ and } 0 < \delta < 1$$

hold. It means that $s(t) = 0$, $\dot{s}(t) = 0$ after a finite time interval, since sum of geometric progression $\sum_{i=1}^{\infty} \Delta t_i$ is bounded. As a result, the second-order sliding mode is enforced in the system with an unknown disturbance d. Similar to the twisting algorithm, implementation of the control needs a time derivative of s; therefore, the relative degree is equal to one as for all types of the conventional sliding mode control.

References

1. Levant, A.: Sliding order and sliding accuracy in sliding mode control. Int. J. Control. **58**(6), 1247–1263 (1993)
2. Shtessel, Y., Edwards, C., Fridman, L., Levant, A.: Sliding Mode Control and Observation. Control Engineering Series. Birkhauser, NY (2014)
3. Loukyanov, A., Utkin, V.: Methods of reducing equations of dynamic systems to a regular form. Autom. Remote. Control. **42**(4), 413–420 (1981)
4. Utkin, V.: Sliding Modes in Control and Optimization. Springer, Berlin (1992)
5. Utkin, V., Guldner, J., Shi, J.: Sliding Mode Control in Electro-Mechanical Systems. CRC Press (2009)
6. Drazenovic, B.: The invariance conditions in variable structure systems. Automatica **5**(3), 287–295 (1969)

7. Atassi, A.N., Khalil, H.K.: Separation results for stabilization of nonlinear systems using different high-gain observer designs. Syst. Control. Lett. **39**, 183–191 (2000)
8. Oh, S., Khalil, H.K.: Nonlinear output feedback tracking using high-gain observer and variable structure control. Automatica **33**, 1845–1856 (1997)
9. Edwards, C., Spurgeon, S.: Sliding Mode Control: Theory and Applications. Taylor and Francis (1998)
10. Poznyak, A.: Advanced Mathematical Tools for Automatic Control Engineers. Deterministic Technique, vol. 1. Elsevier, Amsterdam, Netherlands (2008)
11. Utkin, V.I., Shi, J.: Integral sliding mode in systems operating under uncertainty conditions. In: Proceedings of IEEE CDC, vol. 4, Kobe, Japan, pp. 4591–4596 (1996)
12. Fridman, L., Poznyak, A., Bejarano, F.J.: Robust Output LQ Optimal Control via Integral Sliding Modes. Birkhäuser, Springer Science, Series Systems and Control: Foundations and Applications, New York (2014)
13. Gutman, S.: Uncertain dynamic systems—a lyapunov min-max approach. IEEE Trans. Autom. Control **AC-24**, 437–449 (1979)
14. Gutman, S. Leitmann, G.: Stabilizing feedback control for dynamic systems with bounded uncertainties. In: Confrence on Decision and Control, pp. 94–99 (1976)
15. Orlov, Y., Utkin, V.: Unit sliding mode control in infinite-dimensional systems. J. Appl. Math. Comput. Sci. **8**, 7–20 (1998)
16. Khalil, H.: Nonlinear Systems. Prentice Hall (2002)
17. Levant, A.: Principles of 2-sliding mode control design. Automatica **43**(4), 576–586 (2007)
18. Bhat, S.P., Bernstein, D.S.: Geometric homogeneity with applications to finite-time stability. Math. Control Signals Syst. **17**(2), 101–127 (2005)
19. Utkin, V.: Mechanical energy-based lyapunov function design for twisting and super-twisting sliding mode control. IMA J. Math. Control Inf. **32**(4), 675–688 (2015)
20. Orlov, Y.: Finite time stability and robust control synthesis of uncertain switched systems. SIAM J. Control Optim. **43**(4), 1253–1271 (2005)
21. Bartolini, G., Ferrara, A., Usai, E.: Chattering avoidance by second-order sliding mode control. IEEE Trans. Autom. Control **43**(2), 241–246 (1998)
22. Bartolini, G., Pisano, A., Punta, E., Usai, E.: A survey of application of second-order sliding mode control to mechanical systems. Int. J. Control **76**(9/10), 875–892 (2003)

Chapter 4
Lyapunov Stability Tools for Sliding Modes

Abstract In this chapter, the Lyapunov stability tools for sliding modes analysis are considered. The notion of finite-time stabilization is introduced, and the finite-time suitability analysis is discussed based on the homogeneity property of the considered controllable system. The realization of fixed-time stability is also presented.

Keywords Lyapunov stability tools · Finite-time suitability · Homogeneity · Fixed-time stability

The design methodology of sliding mode control implies the selection of sliding manifold with the desired motion properties, and then selection of the discontinuous control, enforcing the state to reach this manifold, and the existence of sliding mode. Sliding mode is governed by equation with Lipschitz continuous right-hand side, and any design method of the conventional control theory can be applied. The second problem dealing with reaching and existence condition is non-trivial reduced-order stability one due to discontinuities in control. It means that the design methodology for sliding mode control is developed mainly in the context of reaching conditions. The objective of this development is finite-time convergence to the sliding manifold and estimation of the reaching time. For the conventional sliding mode with m-dimensional control and $(n - m)$-dimensional sliding mode equations, the design procedures based on Lyapunov function were demonstrated in the previous section. The above problem becomes more complicated for SOSM, since it should be solved using one-dimensional control input for the nonlinear two-dimensional system. As mentioned above, the lower estimates of reaching time were found in [1, 4] and they can be approached asymptotically. This section demonstrates how the reaching problem with uncertainty conditions can be solved based on Lyapunov functions.

The Lyapunov function method, being initially proposed for smooth dynamic systems, remains a classical analysis tool for Filippov's dynamics in discontinuous systems (see [3–5]), whose trajectories are viewed in the sense of Filippov or according to the equivalent control method. It is well recognized to be useful not only for stability analysis but also for the synthesis of robust sliding motions. In order to analyze sliding mode systems, differentiable Lyapunov functions suffice in many cases. For instance, the Lyapunov analysis of the conventional sliding mode algorithms, demonstrated above, is first applied for the motion beyond the switching

V. Utkin et al., *Road Map for Sliding Mode Control Design*,
SpringerBriefs in Mathematics,
https://doi.org/10.1007/978-3-030-41709-3_4

manifold to establish the attractiveness of this manifold. Once the trajectories are on the switching manifold, smooth dynamics are restored, and the standard Lyapunov theory is in force provided that the SM exists at any point of the manifold.

Recently, non-smooth Lyapunov functions were involved to analyze finite-time stability of the so-called second-order sliding mode (SOSM) systems. In this kind of systems, sliding motions are known to occur only on a discontinuity manifold of co-dimension 2 while the system dynamics cross discontinuity manifolds of co-dimension 1 countably many times within a finite time interval, preceding the SOSM. Available methods of proving finite-time stability of such SM systems are reviewed and illustrated next.

4.1 Finite-Time Stabilization: What is This?

Interest in finite-time stabilization dates back to the early work on optimal control design, presenting the so-called *Fuller effect* [6]. The optimal control for the second-order system

$$\ddot{x}(t) = u(x(t), \dot{x}(t)), \quad t > 0, \ x(t), u(t) \in \mathbb{R}, \quad |u| \le 1 \tag{4.1}$$

with the criterion $\int_0^\infty x^2(t)dt$ is the discontinuous state function

$$u(x, \dot{x}) = -\text{sign}(s(x, \dot{x})), \quad s(x, \dot{x}) = x + c\dot{x}|\dot{x}|. \tag{4.2}$$

The constant parameter $c > 0$ is such that the optimal trajectories cross the switching line $s = 0$ at a countably infinite number of points, located within a finite time interval. The time intervals between successive switches decrease as a geometric progression and result in a finite accumulation point, known as *Zeno behavior*. The existence of a finite accumulation point means finite settling time or finite-time convergence.

The system with SOSM depicted in Fig. 4.1 is identical to the above optimal control system. Both systems are governed by similar second-order equations with parabolic switching lines. For high values of c, the Fuller effect disappears because the sliding mode occurs on a full measure time interval (see Fig. 4.2).

SOSM in twisting and super-twisting systems is different from the above system with the Fuller effect: the conventional sliding mode in the reaching phase for them cannot appear for any parameter combination. At the same time, twisting and super-twisting systems exhibit the properties of the Fuller effect: infinite countable number of switching points at a finite time interval and finite-time convergence.

Fig. 4.1 Process without
sliding mode in the reaching
stage

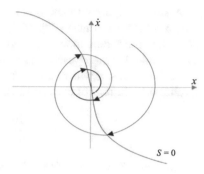

Fig. 4.2 Fuller effect
disappearance

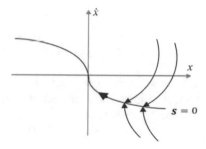

4.2 Strict Lyapunov Functions

Currently, the most popular mathematical technique of proving finite-time stability
of a n-vector system

$$\dot{x}(t) = \phi(t, x(t)) \tag{4.3}$$

with a piecewise continuous (or even locally measurable) right-hand side is finding a
Lyapunov composite function $V(x(t))$ such that its time derivative along the system
trajectories admits estimation

$$\dot{V}(x(t)) \leq -q V^\rho(x(t)) \tag{4.4}$$

for almost all t and for some constants $q > 0, \rho \in [0, 1)$. For such a function, hereafter
referred to as a *strict Lyapunov function*, the expression

$$V(x(t)) = 0$$

holds true (cf. [4, Lemma 4.3]) for all $t \geq T_{ste}$, where

$$T_{ste} = [q(1 - \rho)]^{-1} V^{1-\rho}(x(0)) \tag{4.5}$$

is a settling time estimate. A similar analysis can be provided for non-Lipschitz
Lyapunov functions using a generalized framework (see, e.g., [8–10]).

A strict Lyapunov function is harder to design than an exponentially decaying one, resulting from inequality (4.4) with $\rho = 1$. There are various examples (see [7, 11]) in the sliding mode literature where a strict Lyapunov function is identified. The methodology, proposed in [12], is based on partitioning the state space of (4.3) into subdomains, in which a Lyapunov function $V(x)$ is designed analytically, and it relies on solutions of the first-order partial differential equation

$$\frac{\partial V}{\partial x} \tilde{\phi}(x) = -q V^{\rho}. \tag{4.6}$$

References [13, 14] identify a strict Lyapunov function for the super-twisting system

$$\left.\begin{aligned} \dot{s}_1 &= s_2 - k_1 \sqrt{|s_1|}\, \mathrm{sign}s_1, \\[2mm] \dot{s}_2 &= -k_2 \mathrm{sign}s_1 + w(t, s_1, s_2), \quad k_1, k_2 > 0, \end{aligned}\right\} \tag{4.7}$$

written in terms of the variables s_1, s_2 and affected by uniformly bounded disturbances which additionally admit the dependence on the state variable x and such that the magnitude estimate

$$|w(t, s_1, s_2)| \leq M = \mathrm{const} \tag{4.8}$$

holds almost everywhere. The Lyapunov function for (4.7) is selected as a quadratic form

$$V_{stw}(s_1, s_2) = \zeta^{\mathsf{T}} P \zeta \tag{4.9}$$

with the positive-definite matrix

$$P = \frac{1}{2} \begin{bmatrix} 4k_2 + k_1^2 & -k_1 \\ -k_1 & 2 \end{bmatrix} \tag{4.10}$$

with respect to

$$\zeta^{\mathsf{T}} = \begin{bmatrix} \sqrt{|s_1|}\, \mathrm{sign}s_1, & s_2 \end{bmatrix}. \tag{4.11}$$

Taking into account that no sliding modes occur on the discontinuity surface $x_1 = 0$ but the origin, the time derivative of the Lyapunov function $V_{stw}(x(t))$ is then estimated along the trajectories of (4.7). Provided that

$$k_2 > 3M + 2M^2 k_1^{-2}, \tag{4.12}$$

the desired upper estimate (4.4) is shown to hold true with $\rho = \dfrac{1}{2}, q = \lambda_{min}^{1/2}$ $(P)\lambda_{max}^{-1}(P)\lambda_{min}(Q)$, and

$$Q = \frac{1}{2} \begin{bmatrix} 2k_2 + k_1^2 - M & -k_1 - 2Mk_1^{-1} \\ -k_1 - 2Mk_1^{-1} & 1 \end{bmatrix} > 0. \tag{4.13}$$

Since the above relations are found in a computable form inequality (4.12) can be viewed as a tuning parameter rule, guaranteeing the state convergence to zero in a predetermined finite time given by (4.5).

As shown in [15], specifying the Lyapunov function of the super-twisting dynamics (4.7), (4.8) in the form

$$V_{stw}(x_1, x_2) = 2k_2|x_1| + \frac{1}{2}x_2^2 + \frac{1}{2}\left[e_2 - k_1\sqrt{|x_1|}\,\text{sign}\,x_1\right]^2 \qquad (4.14)$$

yields inequality (4.4) with $\rho = \dfrac{1}{2}$ and with the explicit decay rate

$$q = \sqrt{2k_2}\,\min\left\{\frac{2(k_1k_2 - M - Mk_1)}{3k_1^2 + 4k_2}, \frac{k_1 - 2M}{4}\right\}. \qquad (4.15)$$

A less conservative tuning parameter rule

$$\min\{\frac{k_1}{2}, \frac{k_1k_2}{1 + k_1}\} > M \qquad (4.16)$$

is thus obtained: for instance, given $M \gg 1$, the tuning rule (4.16) requires that $k_2 > M$, whereas (4.12) would require $k_2 > 3M$. Coupled to the settling time estimate (4.5), relations (4.15), (4.16) allow one to tune the super-twisting gains k_1 and k_2 to have any desired settling time. An alternative settling time assignment, which is based on integral HOSM design, can be found in [16].

4.3 Finite-Time Stability and Homogeneity

An alternative to proving finite-time stability through strict Lyapunov functions is via first proving asymptotic stability by identifying a non-strict Lyapunov function [17, 18]) and applying an extension of the invariance principle to discontinuous systems [19]). Once the asymptotic stability is established, the finite-time convergence of trajectories is verified by using weighted homogeneity [20] of system dynamics. The homogeneity concept for a discontinuous system (4.3) is introduced in [21].

Definition 4.1 A piecewise continuous vector function (vector field) $\phi = (\phi_1, \phi_2, \ldots, \phi_n)^\mathsf{T}$ of the variables $t \in \mathbf{R}$ and $x = (x_1, x_2, \ldots, x_n)^\mathsf{T} \in \mathbf{R}^n$ is said to be **locally homogeneous of degree** $\kappa \in \mathbf{R}$ with respect to dilation $\mathbf{r} = (r_1, r_2, \ldots, r_n)^\mathsf{T}$, where $r_i > 0$, $i = 1, 2, \ldots, n$ (as well as system (4.3) is) if there exists a constant $c_0 > 0$ and a ball $B_\delta \in \mathbf{R}^n$ such that

$$\phi_i(c^{-\kappa t}, c^{r_1}x_1, c^{r_2}x_2, \ldots, c^{r_n}x_n) = c^{\kappa + r_i}\phi_i(t, x_1, x_2, \ldots, x_n) \qquad (4.17)$$

for all $c \geq c_0$ and almost all

$$(x_1, x_2, \ldots, x_n)^\mathsf{T} \in B_\delta, \ (c^{r_1} x_1, c^{r_2} x_2, \ldots, c^{r_n} x_n)^\mathsf{T} \in B_\delta,$$

and almost all $t \in \mathbf{R}$. Provided that relation (4.17) is additionally satisfied almost everywhere in $\mathbf{R} \times \mathbf{R}^n$, the function ϕ and system (2.1) are said to be **globally homogeneous** of the same degree and dilation.

Recall that the homogeneity of a Lyapunov function $V(x)$, $x \in \mathbf{R}^n$ is as follows.

Definition 4.2 A scalar continuous function $V(x)$ is called homogeneous of degree $p \in \mathbf{R}$ with respect to dilation $r_1, r_2, \ldots, r_n$, where $r_i > 0$, $i = 1, \ldots, n$, if

$$V\left(c^{r_1} x_1, c^{r_2} x_2, \ldots, c^{r_n} x_n\right) = c^p V(x_1, x_2, \ldots, x_n)$$

for all $(x_1, x_2, \ldots, x_n)^\mathsf{T} \in \mathbf{R}^n$ and for all $c > 0$.

Example 4.1

1. The function $f(x) = x^4$ is homogeneous of degree $p = 4$ and dilatation $r = p = 4$ since
   $$f(cx) = (cx)^4 = c^4 x^4 = c^4 f(x).$$

2. The function $f(x_1, x_2) = x_1^2 + x_2^4$ is homogeneous of degree p and dilatation $(r_1 = p/2, r_2 = p/4)$. Indeed,
   $$f(c^{r_1} x_1, c^{r_2} x_2) = (c^{r_1} x_1)^2 + (c^{r_1} x_1)^4 =$$
   $$c^{2r_1} x_1^2 + c^{4r_2} x_2^4 \overset{c^{2r_1} = c^{4r_2}}{=} c^{2r_1} \left(x_1^2 + x_2^4\right) = c^p f(x_1, x_2)$$
   for dilatations $r_1 = p/2, r_2 = p/4$ and any $p > 0$.

Both the disturbance-free twisting and super-twisting systems (3.16) and (4.7) prove to be globally homogeneous of degree $\kappa = -1$ and *dilation* $\mathbf{r} = (2, 1)$ while straightforwardly validating relation (4.17), properly specified for these systems. It should be noted, however, that their perturbed versions (3.16) and (4.7) are no longer homogeneous because of the presence of the disturbance term $w(t, x_1, x_2)$. These systems may, however, be viewed as quasihomogeneous ones if embedded into the framework of a differential inclusion

$$\ddot{x}(t) \in \Phi(x(t), \dot{x}(t)) + \Psi. \tag{4.18}$$

Then the homogeneity is defined in a similar manner to Definition 4.1, given above. The right-hand side of (4.18) is a multivalued function, composed of the Filippov set $\Phi(x, \dot{x})$ of the discontinuous function

$$\phi(x, \dot{x}) = -\alpha \mathrm{sign} x - \beta \mathrm{sign} \dot{x}$$

and the permanent set

$$\Psi = \{x \in \mathbf{R} : |x| \leq M\},$$

formed by admissible disturbances $w(t, x_1, x_2)$ subject to the rectangular constraint (4.8) on their magnitude. The differential inclusion (4.18), inspected with the embedded twisting algorithm and with the captured class of external disturbances, is shown [7] to be globally homogeneous of the same degree $\kappa = -1$ and dilation $\mathbf{r} = (2, 1)$ as that of the disturbance-free twisting system (4.7). Just in case, such a generator of a homogeneous differential inclusion (4.18) is referred to as a *quasihomogeneous system* of the corresponding degree and dilation.

There is considerable literature employing the homogeneity technique for both continuous and discontinuous vector fields. It is demonstrated in [22, 23] that a globally asymptotically stable continuous homogeneous vector field admits a global homogeneous Lyapunov function, and it is then extended in [24] to discontinuous vector fields. Reference [25] employs the approach of [23] to establish finite-time stability of continuous vector fields when the degree of homogeneity is negative by showing the existence of strict Lyapunov functions. Locally (quasi-) homogeneous functions are well recognized [21] to result in finite-time convergence of discontinuous systems as well when the (quasi-) homogeneity degree is negative [21]. The homogeneous HOSM, proposed in [26], proves finite-time convergence of trajectories also using homogeneity of discontinuous controllers.

The majorant curve technique [27, 28] and homogeneous domination-based approach [29] employ the design of a homogeneous output feedback controller such that the controller stabilizes the nominal linear system in the presence of the uncertainties. The principles of homogeneous domination at zero and at infinity have been introduced originally in [22] [Theorems 6, 7]. Based on the philosophy of homogeneous domination, reference [30] gives a mathematical tool called homogeneity in the bi-limit, which defines a homogeneous approximation. The connection with the homogeneous domination approach is made clear in [30, Sect. 5]. The connection to uniform finite-time convergence of the trajectories of a class of nonlinear autonomous systems is also obtained in [30, Corollary 2.24, Example 5.5] using the concept of homogeneity in the bi-limit.

A philosophically similar concept of quasihomogeneity, concerning discontinuous systems, leads to the uniform finite-time stability of the origin in the presence of bounded persistent external disturbances, resulting in the equiuniform finite-time stability [4, Sect. 4.4]. Reference [21] is the first article where such equiuniform finite-time stability is investigated under uniformly bounded persistent external disturbances. An equiuniform settling time estimate is obtained with respect to the entire class of such disturbances. The proposed equiuniform finite-time stability analysis is based on the premise of equiuniform asymptotic stability and is illustrated by the PD-augmented perturbed twisting system

$$\left. \begin{aligned} \dot{s}_1 &= s_2, \\ \dot{s}_2 &= -\alpha\mathrm{sign}(s_1) - \beta\mathrm{sign}(s_2) - hs_1 - ps_2 + \omega \end{aligned} \right\} \tag{4.19}$$

with the gains $h, p > 0$ [21].

First, equiuniform stability of (4.19) is verified with the non-strict Lyapunov function

$$\tilde{V}(s_1, s_2) = \alpha|s_1| + \frac{1}{2}(s_2^2 + hs_1^2). \tag{4.20}$$

Indeed, differentiating (4.20) along the perturbed system (4.19), initialized outside the origin, yields the equiuniform estimate

$$\dot{\tilde{V}}(s_1(t), s_2(t)) \leq -(\beta - M)|s_2(t)| \tag{4.21}$$

for almost all t regardless of whichever disturbance (4.8) affects the system. While deriving (4.21), it has been taken into account that no sliding modes occur along the discontinuity lines $x_1 = 0$ and $x_2 = 0$ except the origin $x_1 = 0$ and $x_2 = 0$.

Next, a family of semiglobal Lyapunov functions

$$V_R(s_1, s_2) = \tilde{V}(s_1, s_2) + \mu_R s_1 s_2 = \alpha|x_1| + \frac{1}{2}(s_2^2 + hs_1^2) + \mu_R s_1 s_2, \tag{4.22}$$

parameterized with $R > 0$, is involved with sufficiently small positive weight parameters such that each Lyapunov function (4.22) is positive definite on the corresponding compact set

$$D_R = \{(s_1, s_2) \in \mathbf{R}^2 : \tilde{V}(s_1, s_2) \leq R\}. \tag{4.23}$$

The parameter R defines the domain of the Lyapunov representative V_R. The higher the value of R, the smaller the value of the cross-term weight μ.

Since due to (4.21), all trajectories of the perturbed system (4.19), initialized within compact (4.23), cannot escape from this compact, the estimate of the time derivative of (4.22) along these trajectories is obtained in the form

$$\dot{V}_R(s_1(t), s_2(t)) \leq -K_R V_R(s_1(t), s_2(t)) \tag{4.24}$$

with some positive equiuniform exponential decay rate K_R on compact (4.23). Certainly, the exponential decay (4.24) on compact (4.23) does not lead to the exponential stability of the perturbed system (4.19) because the decay rate $K_R \to 0$ as $R \to \infty$, but it does guarantee the global equiuniform asymptotic stability of (4.19).

Finally, the equiuniform finite-time convergence is established based on the local homogeneity of (4.19). Taking into account that the PD+disturbance term

$$-hx_1 - px_2 + \omega(t, x_1, x_2)$$

remains locally uniformly bounded for admissible disturbances (4.8), system (4.19) is embedded into the differential inclusion framework (4.18). This allows one to ensure that the system is indeed locally (but not globally) quasihomogeneous of the same degree $\kappa = -1$ and dilation $\mathbf{r} = (2, 1)$ as that of the disturbance-free twisting system (4.19).

Thus, the PD-augmented perturbed twisting system (4.19) is globally equiuniformly asymptotically stable and it is locally quasihomogeneous of degree $\kappa < 0$.

By applying the quasihomogeneity principle [4, Theorem 4.1], the global equiuniform finite-time stability of (4.19) is established.

In addition, the quasihomogeneity technique has successfully been applied in [31] for tuning the twisting controller parameters to impose a predetermined settling time estimate on the closed-loop system.

The homogeneous HOSM, proposed in [26], proves finite-time convergence of trajectories also using homogeneity of discontinuous controllers.

4.4 Fixed-Time Stability

The notion of the so-called fixed-time stability refers to the case of finite-time stability when the finite settling time can be specified independent of the initial conditions. The fixed-time stability can be established in terms of a strict Lyapunov function V that along the solutions of (2.1) satisfies a stronger version

$$\dot{V}(x(t)) \leq -\eta(V(x(t))) \qquad (4.25)$$

of the Lyapunov derivative estimate (4.4) with a right-hand side such that the ODE

$$\dot{V}(x(t)) = -\eta(V(x(t)))$$

possesses fixed-time stable solutions. An example of such a function is

$$\eta(V) = \sqrt{V + V^3}$$

Polyakov [32]. Indeed, the corresponding equation

$$\dot{V}(x(t)) = -\sqrt{V + V^3} \qquad (4.26)$$

possesses solutions with the convergence rate higher than that of

$$\dot{V}(x(t)) = -\sqrt{V^3} \qquad (4.27)$$

beyond the level set $V = 1$ and higher than that of

$$\dot{V}(x(t)) = -\sqrt{V} \qquad (4.28)$$

within this level set. Since the solutions

$$V(t) = 4V(0)(2 + t\sqrt{V(0)})^{-2}$$

of (4.27) with initial conditions $V(0) > 1$ reach the level set $V = 1$ in

$$T(V(0)) = 2(\sqrt{V(0)} - 1)/\sqrt{V(0)} \le 2, \tag{4.29}$$

whereas the solutions to (4.28), initialized within $V(0) \le 1$ and being a particular case of (4.4), escape to zero according to (4.29) at most at $T = 2$, the solutions to (4.26) escape to zero at most $T = 4$, regardless of initial conditions. Thus, (4.26) is actually fixed-time stable.

For continuous vector fields, fixed-time stability has been derived in the context of systems homogeneous in the bi-limit [30, Corollary 2.24]. The fixed-time stability of a discontinuous super-twisting-based differentiator was proved in [33], where it was shown that the predefined convergence time does not depend on the initial condition. A more recent result in [32] proposes new nonlinear controllers for fixed-time stabilization of linear control systems. The most recent reference in this area can be found in [34], which studied the properties of the settling time function resulting from the application of finite-time convergent sliding modes, giving rise to Filippov inclusions. An important limitation of the fixed-time stable discontinuous systems under forward Euler discretization was established in [34, Theorem 3]. It is shown that there exist solutions of the inclusions whose norm escapes to infinity. Note that the implicit Euler discretization does not have this limitation and preserves the fixed-time convergence rate [35].

References

1. Utkin, V., Guldner, J., Shi, J.: Sliding Mode Control in Electro-Mechanical Systems. CRC Press (2009)
2. Khalil, H.: Nonlinear Systems. Prentice Hall (2002)
3. Utkin, V.: Sliding Modes in Control and Optimization. Springer, Berlin (1992)
4. Orlov, Y.: Discontinuous Systems: Lyapunov Analysis and Robust Synthesis Under Uncertainty Conditions. Springer, Berlin (2008)
5. Yakubovich, V., Leonov, G., Gelig, A.: Stability of Stationary Sets in Control Systems with Discontinuous Nonlinearities. World Scientfic (2004)
6. Fuller, A.: Relay control systems optimized for various performance criteria. In: Proceedings of the 1st IFAC Triennial World Congress, pp. 510–519 (1960)
7. Orlov, Y.: Finite time stability and robust control synthesis of uncertain switched systems. SIAM J. Control Optim. **43**(4), 1253–1271 (2005)
8. Roxin, E.: On finite stability in control systems. Rendiconti del Circolo Matematico di Palermo **15**(3), 273–283 (1966)
9. Chellaboina, V., Leonessa, A., Haddad, W.: Generalized Lyapunov and invariant set theorems for nonlinear dynamical systems. Syst. Control Lett. **38**, 289–295 (1999)
10. Polyakov, A., Fridman, L.: Stability notions and Lyapunov functions for sliding mode control systems. J. Franklin Inst. **351**(4), 1831–1865 (2014)
11. Kamal, S., Moreno, J., Chalanga, A., Bandyopadhyay, B., Fridman, L.: Continuous terminal sliding-mode controller. Automatica **69**, 308–314 (2016)
12. Polyakov, A., Poznyak, A.: Reaching time estimation for super-twisting second order sliding mode controller via lyapunov function designing. IEEE Transactions on Automatic Control **54**(8), 1951–1955 (2009)
13. Moreno, J., Osorio, M.: A lyapunov approach to second-order sliding mode controllers and observers. In: 47th IEEE Conference on Decision and Control (CDC), pp 2856–2861 (2008)

14. Moreno, J., Osorio, M.: Strict lyapunov functions for the super-twisting algorithm. IEEE Trans. Autom. Control **57**(4), 1035–1040 (2012)
15. Orlov, Y., Aoustin, Y., Chevellereau, C.: Finite time stabilization of a perturbed double integrator - part i: continuous sliding mode-based output feedback synthesis. IEEE Trans. Autom. Control **56**(3), 1035–1040 (2010)
16. Levant, A., Alelishvili, L.: Integral high order sliding modes. IEEE Trans. Autom. Control **52**(7), 1278–1282 (2007)
17. Pyatnitskii, E.S.: Control of mechanical systems under uncertainty conditions in the absence of quantitative information on the current state. Autom. Remote. Control **60**(5), 739–743 (1999)
18. Orlov, Y.: Extended invariance principle for nonautonomous switched systems. IEEE Trans. Autom. Control **48**(5), 1448–1552 (2003a)
19. Alvarez, J., Orlov, Y., Acho, L.: An invariance principle for discontinuous dynamic systems with application to a coulomb friction oscillator. ASME J. Dyn. Syst., Meas., Control **122**(4), 687–690 (2000)
20. Zubov, V.I.: Methods of A.M. Lyapunov and Their Applications. Noordhoff, Leiden (1964)
21. Orlov, Y.: Finite time stability of homogeneous switched systems. In: 42nd IEEE Conference on Decision and Control (CDC), pp. 4271–4276 (2003b)
22. Zubov, V.I.: On systems of ordinary differential equations with generalized homogenous right-hand sides. Izvestia vuzov. Mathematica. **1**, 80–88 (1958) (in Russian)
23. Rosier, L.: Homogeneous lyapunov function for homogeneous continuous vector field. Syst. Control Lett. **19**(6), 467–473 (1992a)
24. Rosier, L.: Inverse of lyapunov's second theorem for measurable functions. In: Proceedings of IFAC Symposium on Nonlinear Control Systems (NOLCOS), pp. 2856–2861 (1992b)
25. Bhat, S.P., Bernstein, D.S.: Geometric homogeneity with applications to finite-time stability. Math. Control Signals Syst. **17**(2), 101–127 (2005)
26. Levant, A.: Homogeneity approach to high-order sliding mode design. Automatica **41**(5), 467–473 (2005)
27. Levant, A.: Sliding order and sliding accuracy in sliding mode control. Int. J. Control **58**(6), 1247–1263 (1993)
28. Orlov, Y., Aguilar, L., Cadiou, J.C.: Switched chattering control versus backlash/friction phenomena in electrical servo-motors. Int. J. Control **76**(9/10), 959–967 (2003)
29. Qian, C.: A homogeneous domination approach for global output feedback stabilization of a class of nonlinear systems. In: Proceedings of American Control Conference (ACC), pp. 4708–4715 (2005)
30. Andrieu, V., Praly, L., Astolfi, A.: Homogeneous approximation, recursive observer design, and output feedback. SIAM J. Control Optim. **47**(4), 1814–1850 (2008)
31. Oza, H., Orlov, Y., Spurgeon, S.: Continuous uniform finite time stabilization of planar controllable systems. SIAM J. Control Optim. **53**(3), 1154–1181 (2015)
32. Polyakov, A.: Nonlinear feedback design for fixed-time stabilization of linear control systems. IEEE Trans. Autom. Control **57**(8), 2106–2110 (2012)
33. Cruz-Zavala, E., Moreno, J., Fridman, L.: Uniform robust exact differentiator. IEEE Trans. Autom. Control **56**(11), 2727–2733 (2011)
34. Levant, A.: On fixed and finite time stability in sliding mode control. In: 52nd IEEE Conference on Decision and Control (CDC), pp. 4260–4265 (2013)
35. Polyakov, A., Efimov, D., Brogliato, B.: Consistent discretization of finite-time and fixed-time stable systems. SIAM J. Control Optim. **57**(1), 78–103 (2019)

Chapter 5
SM Observers

Abstract Most of the design methods considered above were developed under the assumption that the state vector is available. In practice, however, only a part of its components may be measured directly. The output feedback SMC method is applicable to rather limited types of systems. An alternative approach is designing asymptotic observers, which are dynamic systems for estimating all components of the state vector using those measured directly. Sliding mode modifications for state observation Chap. 14, [2, Chap. 6], and [3] are presented in this section. Time derivatives are needed for design in canonical space and for high-order sliding mode control. Differentiation design problem is discussed as well.

Keywords Observers for linear time-invariant systems · Uncertain systems · Differentiators

5.1 Observers for Linear Time-Invariant Systems

The conventional asymptotic state observer for LTI observable system

$$\dot{x}(t) = Ax(t) + Bu(t), \quad y(t) = Cx(t), \\ x(t) \in R^n, \; u(t) \in R^m, \; y(t) \in R^l, \; l < n, \; \text{rank}(C) = l$$

has the same structure as the system under control with an additional input depending on the mismatch between the output $y(t)$ and its estimate

$$\dot{\hat{x}}(t) = A\hat{x}(t) + Bu(t) + L(y(t) - C\hat{x}(t)), \quad \hat{x}(t) \in R^n. \tag{5.1}$$

Matrix L can be selected such that the state estimate $\hat{x}(t)$ tends to the system state $x(t)$ at the desired rate. Let us proceed to the design of a state observer with inputs as discontinuous functions of mismatches. Taking into account that the state vector can be partitioned into two subvectors

$$x^T = (x_1^T, x_2^T)^T, x_1 \in R^{n-l}, x_2 \in R^l$$

© The Author(s), under exclusive license to Springer Nature Switzerland AG 2020
V. Utkin et al., *Road Map for Sliding Mode Control Design*,
SpringerBriefs in Mathematics,
https://doi.org/10.1007/978-3-030-41709-3_5

such that

$$s = C_1 x_1 + C_2 x_2, \det(C_2) \neq 0,$$

write down the observer equation with respect to estimates $\hat{x}_1$ and $\hat{y}$

$$\left. \begin{array}{l} \dot{\hat{x}}_1(t) = A_{11}\hat{x}_1(t) + A_{12}\hat{y}(t) + B_1 u(t) + Lv(t), \\ \dot{\hat{y}}(t) = A_{21}\hat{x}_1(t) + A_{22}\hat{y}(t) + B_2 u(t) - v(t) \\ v(t) = M \, \text{sign}(\hat{y}(t) - y(t)), \, M > 0, \end{array} \right\}$$

where all matrices are constant and are derived from the transformation of the observer Eq. (5.1). The discontinuous vector function $v(t)$ is chosen such that sliding mode is enforced in the manifold

$$\overline{y}(t) = \hat{y}(t) - y(t)$$

and the mismatch $\overline{y}(t)$ between the output $y(t)$ and its estimate $\hat{y}(t)$ is reduced to zero. Matrix L must be found such that the mismatch $\overline{x}_1(t) = \hat{x}_1(t) - x_1(t)$ decays at the desired rate. Sliding mode with $\overline{y}(t) = 0$ occurs in the system with respect to $\overline{x}_1$ and $\overline{y}$

$$\dot{\overline{x}}_1(t) = A_{11}\overline{x}_1(t) + A_{12}\overline{y}(t) + Lv(t),$$
$$\dot{\overline{y}}(t) = A_{21}\overline{x}_1(t) + A_{22}\overline{y}(t) - v(t)$$

with high enough value of $M > 0$. Hence, for bounded initial conditions, sliding mode can be enforced in manifold $\overline{y} = 0$. To derive the sliding mode equation, the solution

$$v_{eq}(t) = A_{21}\overline{x}_1(t)$$

to equation $\dot{\overline{y}}(t) = 0$ with respect to $v(t)$ should be substituted into the first equation

$$\dot{\overline{x}}_1(t) = (A_{11} + LA_{21})\overline{x}_1(t).$$

The desired rate of convergence of $\overline{x}_1(t)$ to zero and convergence of $\hat{x}_1(t)$ to $x_1(t)$ can be provided by a proper choice of matrix L [1] and then

$$x_2(t) = C_2^{-1}(y(t) - C_1 x_1(t)).$$

The observer with the input as a discontinuous function of the output in sliding mode is equivalent to the linear reduced-order Luenberger observer [1]. However, if the plant and observed signal are affected by noises, the nonlinear observer is preferable, since it exhibits adaptivity properties with respect to varying parameters of the noises [4].

5.2 Observers for Uncertain Systems

5.2.1 Second-Order System

The main ideas of designing observers for nonlinear uncertain systems can be demonstrated for the second-order system:

$$\dot{x}(t) = Ax(t) + df(t, x(t), u(t)) + bu(t), \quad y(t) = cx(t),$$

where $x(t) \in R^2$, $y(t) \in R$, A, c, d, b are constant matrix and vectors, f is a scalar bounded disturbance $\|f\| < f_0$ where f_0 is constant. The pair (A, c) is assumed to be observable.

The observable system can be transformed into the form [1]

$$\left.\begin{aligned}
\dot{x}_1 &= a_{11}x_1 + a_{12}y + d_1 f + b_1 u, \quad a_{11} < 0. \\
\dot{y} &= a_{21}x_1 + a_{22}y + d_2 f + b_2 u.
\end{aligned}\right\}$$

The observer

$$\left.\begin{aligned}
\tfrac{d}{dt}\hat{x}_1 &= a_{11}\hat{x}_1 + a_{12}y + b_1 u, \\
\tfrac{d}{dt}\hat{y} &= a_{21}\hat{x}_1 + a_{22}y + b_2 u + v, \\
v &= -v_0 \text{sign}(\hat{y} - y), \ v_0 > |d_2| f_0
\end{aligned}\right\}$$

is designed, copying the parameters of the above system and assuming that $d_1 = 0$. The error equation has the form

$$\left.\begin{aligned}
\dot{e}_x &= a_{11}e_x, \\
\dot{e}_y &= a_{21}e_y + d_2 f + v.
\end{aligned}\right\}$$

The mismatch $e_x = \hat{x}_1 - x_1$ tends to zero ($a_{11} < 0$) and component x_1 of the state vector is found. Discontinuous input v enforces sliding mode with

$$e_y = \hat{y} - y = 0$$

and

$$v_{eq} = d_2 f,$$

which can be obtained using a low-pass filter [5, Chap. 2]. It means that the observer can be applied for finding both the state vector and disturbance or for fault detection if f models the effect of fault.

5.2.2   Uncertain System of an Arbitrary Order

The above method can be easily generalized for the uncertain system of an arbitrary order

$$\dot{x} = Ax + Bu + Df(t, x, u), \quad y = Cx, \tag{5.2}$$

where $x \in R^n, u \in R^m\ y \in R^p, A, C, D, B$ are constant matrices, $f(t, x, u) \in R^q$ is a bounded disturbance. The pair (A, C) is assumed to be observable [6]. The observable system can be transformed into the form

$$\left.\begin{array}{l} \dot{x}_1 = A_{11}x_1 + A_{12}y + D_1 f + B_1 u, \\[2mm] \dot{y} = A_{21}x_1 + A_{22}y + D_2 f + B_2 u, \end{array}\right\}$$

where all matrices are derived from (5.2) and A_{11} is a Hurwitz matrix (like $a_{11} < 0$). Similar to the assumptions above $D_1 = 0$ (like $d_1 = 0$). Then the disturbance and input v in the mismatch equations satisfy the matching conditions (see Sect. 3.5); therefore, the motion in sliding mode does not depend on disturbance.

The equations of the observer and mismatches are of form

$$\left.\begin{array}{l} \dfrac{d}{dt}\hat{x}_1 = A_{11}\hat{x}_1 + A_{12}y + B_1 u, \\[3mm] \dfrac{d}{dt}\hat{y} = A_{21}\hat{x}_1 + A_{22}y + B_2 u + v, \\[3mm] \qquad v = -v_0 \mathrm{sign}(e_y), \end{array}\right\}$$

and

$$\left.\begin{array}{l} \dot{e}_x = A_{11}e_x, \\ \dot{e}_y = A_{21}e_y + D_2 f + v. \end{array}\right\}$$

As for the second-order system vector x_1 and the disturbance can be found. The state and disturbance observer is designed in terms of the original system (A, B, C, D) in [7, 8].

Further developments of the discussed methodology for state and fault estimations can be found in [9, 10] for systems with noise in measurement [11], for full- and reduced-order observers [12], for a set of interconnected systems [13], and for systems with high-order sliding modes [14].

Of course the condition $D_1 = 0$ is rather restrictive. It is not applicable for widely used systems in canonical form

$$\left.\begin{array}{l} \dot{x}_i = x_{i+1}, \ i = 1, \ldots, n - 1, \\ \dot{x}_n = f(t, x, u), y = x_1). \end{array}\right\}$$

The method does not work for our second-order example if

$$d_1 \neq 0, d_2 = 0.$$

The design idea for the general case when condition $D_1 = 0$ does not hold stems from the so-called high-gain observer with specially selected hierarchy of gains tending to infinity (discussed in section "Invariance").

It is interesting that the canonical form for designing a high-gain observer [15] is not obligatory, as shown in the paper, published in Russian [16]. The design procedures for high-gain and sliding mode observers are outlined in this section. Preliminarily demonstrate the idea of high-gain observer design for an arbitrary LTI system. The first equation of the observer for the system in canonical form

$$\dot{x}_i = x_{i+1}, i = 1, \ldots, n-1,$$
$$\dot{x}_n = F(x_1, \ldots, x_n, t) + u,$$
$$y = x_1$$

with respect to mismatch $\bar{x}_i := \hat{x}_i - x_i$ is

$$\frac{d}{dt}\bar{x}_1 := \bar{x}_2 - k_1\bar{x}_1.$$

If $k_1 \to \infty$, then $k_1\bar{x}_1$ tends to $\bar{x}_2$. Similarly, known x_2 can be used to find x_3 with corresponding hierarchy of high gains and so on. Now assume that $y = x_1$ is not scalar but output vector with equations $\dot{x}_1 = Ax$, where x is a state vector and

$$\frac{d}{dt}\bar{x}_1 := A\bar{x} - k_1\bar{x}_1.$$

If scalar gain k_1 tends to infinity, then $k_1\bar{x}_1$ tends to $A\bar{x}$. Similarly, vector $y_1 = A\bar{x}$ can be used to find the next subvector of the state again with corresponding hierarchy of the observer gains.

Now describe in details the high-gain observer design. Any LTI observable system

$$\dot{x} = Ax + Bu, y = Cx, \quad C \text{ is a full rank matrix}$$

can be represented in the form

$$\dot{y} = A_{11}y + A_{12}x_1 + B_1u,$$
$$\dot{x}_1 = A_{21}y + A_{22}x_1 + B_2u,$$

where the pair (A_{22}, A_{12}) is observable. If gain $k_0 \to \infty$ in the observer

$$\dot{\hat{y}} = A_{11}\hat{y} + A_{12}\hat{x}_1 + B_1u + v_0, \quad v_0 = k_0(y - \hat{y}),$$
$$\dot{\hat{x}}_1 = A_{21}\hat{y} + A_{22}\hat{x}_1 + B_2u + v_{x_1},$$

then $\hat{y} \to y$ and $v_0 \to A_{12}(\hat{x}_1 - x_1)$, rank $A_{12} > 0$ (otherwise, the system is not observable), where v_{x_1} to be defined next. Matrix R exists such that $C_1 = RA_{12}$ is a full rank matrix. Then

$$y_1 = R(A_{12}\hat{x}_1 - v_0) = RA_{12}x_1 = C_1x_1.$$

If $\dim(y_1) \geq \dim(x_1)$, then x_1 can be found from y_1 and x can be found from $y = Cx$ and x_1. It is important that the equation for x_1 is not needed for this case. The equation for which the output vector $y_1 = C_1x_1$ can be handled similar to the original system if $\dim(y_1) < \dim(x_1)$. It can be partitioned into two equations with respect to y_1 and new variable x_2 with new output vector $y_2 = C_2x$ and v_{x_2} instead of v_{x_1}. Again vector y_2 can be obtained, if $v_1 = k_1(y_1 - \hat{y}_1)$, $k_1 \to +\infty$, $k_0/k_1 \to \infty$.

After a finite number of steps k with the hierarchy of gains $k_{k-1} \to \infty$, $\dfrac{k_i}{k_{i+1} \cdot \ldots \cdot k_{k-1}} \to \infty$ the condition $\dim(y_k) \geq \dim(x_k)$ holds and x_k can be found from the algebraic equation $y_k = C_kx_k$. All coordinate transformations are non-singular, and the remaining components of the state vector can be calculated from algebraic equations $y = Cx$, $y_i = C_ix_i$, $i = 1, \ldots, k - 1$. The last equations for x_i is not needed; therefore, its right-hand side may depend on any types of uncertainties (unknown parameters, disturbances, nonlinear functions) similar to the systems in canonical form with uncertainties in the last equation.

The high-gain design method is generalized for observers with finite discontinuous inputs. Input $v_0 = k_0(y - \hat{y})$ is replaced by $v_0 = M_0 \operatorname{sign}(y - \hat{y})$ with finite gain M_0. After sliding mode occurs, the average value of v_0 is equal to $A_{12}(\hat{x}_1 - x_1)$ as well and can be obtained using a low-pass filter. Similarly, v_1 is replaced by $v_1 = M_1 \operatorname{sign}(y_1 - \hat{y}_1)$ and so on. All state components can be found in the observer with appropriate hierarchy of low-pass filter time constants.

Similar multistep procedure is applicable for both linear and nonlinear systems [17, 18]. This was demonstrated for time-varying linear systems in [5, Chap. 6].

5.2.3 Analysis of the State Estimation Error Convergence for Matched Disturbances

The observer should be designed for the uncertain system

$$\begin{aligned}
\dot{x}(t) &= Ax(t) + Bu(t) + f(t, x(t), u(t)), \quad y(t) = Cx(t), \\
x(t) &\in R^n, \ u(t) \in R^m, \ y(t) \in R^p, \ \operatorname{rank}(C) = p < n,
\end{aligned} \right\}$$

such that the mismatch $e(t) = \hat{x}(t) - x(t)$ between the estimate and system state is reduced to zero. The disturbance is assumed to satisfy the matching condition

$$\begin{aligned}
f(t, x(t), u(t)) &= B\xi(t, x(t), u(t)), \\
\|\xi(t, x, u)\| &\leq k_1\|u\| + k_2\|y\| + \alpha_0, \ y = Cx,
\end{aligned}$$

where k_1, k_2, α_0 are constants and $u \in R^m$, $x \in R^n$, $y \in R^p$. If the pair (A, C) is observable, then there exists matrix L such that matrix $A + LC$ has stable eigenvalues, i.e., for any $Q > 0$ the solution $P > 0$ to Lyapunov equation

$$(A + LC)^T P + P(A + LC) = -Q$$

can be found.

Assumption. There exist matrices $P, Q \in R^{n \times n}$, $L \in R^{n \times p}$, $F \in R^{m \times p}$ such that $C^T F^T = PB$.

If this condition holds, then the amplitude of discontinuous control can be equal to any upper estimate of the disturbance.

The observer equations [2] are selected with discontinuous input in the form of unit control, introduced in Chapter IV

$$\left.\begin{aligned}
\dot{\hat{x}}(t) &= A\hat{x}(t) + Bu(t) + L(y(t) - C\hat{x}(t)) + v(t), \\
v(t) &= -\rho(u(t), y(t))P^{-1}C^T F^T \frac{F(y(t) - C\hat{x}(t))}{\|F(y(t) - C\hat{x}(t))\|}, \\
\rho(u, y) &= k_1\|u\| + k_2\|y\| + k_0, \quad k_0 > \alpha_0.
\end{aligned}\right\}$$

Calculate time derivative of

$$V(e(t)) = e^T(t)Pe(t)$$

on the trajectories of the system

$$\dot{e}(t) = (A + LC)e(t) - B\xi(t, x(t), u(t)) + v(t).$$

We have

$$\left.\begin{aligned}
\dot{V}(e(t)) &\leq -e^T(t)Qe(t) + 2e^T C^T F^T \xi(t, x(t), u(t)) \\
&\quad -2\rho(u(t), y(t))e^T(t)C^T F^T \frac{FCe(t)}{\|FCe(t)\|} \leq \\
&\quad -e^T(t)Qe(t) - 2(k_0 - \alpha_0)\|FCe(t)\| < 0.
\end{aligned}\right\}$$

It means that the mismatch tends to zero, and the estimate tends to the state vector. The design method is applicable under structural constraint

$$C^T F^T = PB$$

and for disturbance, depending on the control and output only. If the disturbance depends on the state vector, then convergence of the mismatch to zero can be guaranteed for a bounded domain of initial conditions similar to the observer in the previous section.

5.2.4 Differentiators

5.2.4.1 Single Differentiation

Implementation of sliding mode control both for systems in canonical space of 60s and for systems with high-order sliding mode [19, Chap. 6] needs time derivatives of different orders. During the entire history of sliding mode control theory, the researchers tried to use sliding mode for designing differentiators. The simplest one is shown in Fig. 5.1

$$\dot{s}(t) = \dot{f}(t) - v(t), \quad v(t) = M \operatorname{sign}(s(t)),$$

where $s(t) = f(t) - x(t)$. If $M > |\dot{f}|$, then sliding mode occurs with $x(t) \equiv f(t)$. It means that the input of the integrator should be close to time derivative of $f(t)$, but the input $v(t)$ is a discontinuous function. Explanation of this discrepancy is simple: signal $v(t)$ consists of high-frequency and average components. The high-frequency one is filtered out by the integrator, while the average component is equal to time derivative of $f(t)$. The first-order low-pass filter is added to filter out the high-frequency component and its output $z(t)$ should be close to $\dot{f}(t)$. This heuristic motivation can be confirmed analytically.

The block diagram of the two-loop differentiator for estimation of the first and second time derivatives of function $f(t)$ is shown in Fig. 5.2. The procedure of selection of parameters τ, τ_1, μ and upper estimates of differentiation errors can be found in [20].

5.2.4.2 Multiple Differentiation

The problem of multiple differentiation can be reformulated in terms of the dynamic systems governed by equations

$$\dot{x}_i(t) = x_{i+1}(t), i = 1, \ldots, n - 1$$

Functions $x_i, i = 2, \ldots, n$ should be found if

$$f(t) = x_1$$

Fig. 5.1 Block diagram of the first-order sliding mode differentiator

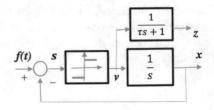

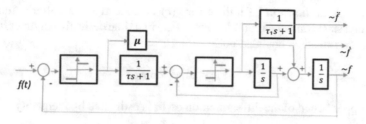

Fig. 5.2 Block diagram of the second-order sliding mode differentiator

is measured. This problem is a particular case of designing an observer for a linear system

$$\dot{x}(t) = Ax(t) + Bu(t), \, y = Cx.$$

As discussed in the previous section, first the high-gain observer is designed with appropriate hierarchy of gains. Then, high gains are replaced by discontinuous functions with low-pass filters and appropriate hierarchy of the low-pass filter time constants. The multistep methodology, offered in [16], is applied directly to the above-formulated problem. After sliding mode occurs in the system

$$\left. \begin{aligned} \dot{\hat{x}}_1(t) &= v_1(t), \\ v_1(t) &= -M \, \mathrm{sign}(\hat{x}_1(t) - x_1(t)), \\ \tau_1 \dot{z}_1(t) + z_1(t) &= v_1(t), \end{aligned} \right\}$$

z_1 is approximately equal to x_2 the time derivative of x_1. The second derivative can be estimated in the same way

$$\left. \begin{aligned} \dot{\hat{x}}_2(t) &= v_2(t), \\ v_2(t) &= -M \, \mathrm{sign}(\hat{x}_2(t) - v_1(t)), \\ \tau_2 \dot{z}_2(t) + z_2(t) &= v_2(t), \end{aligned} \right\}$$

z_2 is approximately equal to x_3 being the second time derivative of x_1. On the next steps, the 3rd,...,$(n-1)$th derivative can be estimated. The upper bounds of the estimation errors can be found in [16].

The design method for super-twisting control can be utilized for the differentiation problem as well [21]. To differentiate time function $f(t)$, the super-twisting controller is designed to reduce the error $s(t)$ between its input $f(t)$ and output $x(t)$ to zero, $s = x - f$:

$$\left. \begin{aligned} \dot{x}(t) &= -\alpha\sqrt{|s(t)|}\mathrm{sign}(s(t)) + y(t), \\ \dot{y}(t) &= -M \, \mathrm{sign}(s(t)) \end{aligned} \right\}$$

with

$$|\ddot{f}(t)| < F_0 = \mathrm{const}$$

and $M > F_0$. As shown in [19], the error $s(t)$ is reduced to zero after a finite time interval and state component $y(t)$ is equal to the first time derivative of function $f(t)$. If $f(t)$ is corrupted by bounded noise

$$|s(t)| \leq \Delta = \text{const},$$

then an upper bound of the differentiation error is estimated by inequality

$$|y(t) - \dot{f}(t)| \leq \alpha_1 \Delta + \alpha_2 \sqrt{\Delta},$$

where α_1 and α_2 are constant coefficients. The average value of

$$\dot{y}(t) = -M \text{ sign}(s(t))$$

is close to second derivative of $f(t)$ and it can be derived as output $z(t)$ of low-pass filter

$$\tau \dot{z}(t) + z(t) = -M \text{ sign}(s(t)).$$

References

1. Utkin, V.: Sliding Modes in Control and Optimization. Springer, Berlin (1992)
2. Edwards, C., Spurgeon, S.: Sliding Mode Control: Theory and Applications. Taylor and Francis (1998)
3. Spurgeon, S.: Sliding mode observers: a survey. Int. J. Syst. Sci. **39**(8), 751–764 (2008)
4. Drakunov, S.V.: On adaptive quasioptimal filter with discontinuous parameters. Autom. Remote Control **44**(9), 1167–1175 (1983)
5. Utkin, V., Guldner, J., Shi, J.: Sliding Mode Control in Electro-Mechanical Systems. CRC Press (2009)
6. Edwards, Christopher, Spurgeon, Sarah K., Patton, Ron J.: Sliding mode observers for fault detection and isolation. Automatica **36**(4), 541–553 (2000)
7. Walcott, B.L., Zak, S.H.: Observation of dynamical systems in the presence of bounded non-linearities/uncertainties. In: Confrence on Decision and Control, pp. 961–966 (1986)
8. Corless, M., Tu, J.: State and input estimation for a class of uncertain systems. Automatica **34**(6), 757–764 (1998)
9. Edwards, C., Spurgeon, S.: On the development of discontinuous observers. Int. J. Control **59**(5), 1211–1229 (1994)
10. Kalsi, K., Hui, S., Zak, S.H.: Unknown input and sensor fault estimation using sliding-mode observers. In: American Control Conference, pp. 1364–1369 (2011)
11. Yang, J., Zhu, F., Wang, X., Xuhui, B.: Robust sliding-mode observer-based sensor fault estimation, actuator fault detection and isolation for uncertain nonlinear systems. Int. J. Control Autom. Syst. **13**(5), 1037–1046 (2015)
12. Dhahri, S., Sellami, A., Hmida, F.B.: Fault detection and reconstruction via both full-order and reduced-order sliding mode observers for uncertain nonlinear systems. In: International Conference on Modelling, Identification and Contro (2015)
13. Menon, P.P., Edwards, C.: Fault estimation using relative information for a formation of dynamical systems. In: Confrence on Decision and Control (2010)

14. Fridman, L., Levant, A., Davila, J.: Observation of linear systems with unknown inputs via high-order sliding-modes. Int. J. Syst. Sci. **38**(10), 773–791 (2008)
15. Oh, S., Khalil, H.K.: Nonlinear output feedback tracking using high-gain observer and variable structure control. Automatica **33**, 1845–1856 (1997)
16. Utkin, V.I.: Control problems for multi-connected plants. In: Motion Separation For Designing State Observer, pp. 91–97. Nauka, Moscow (1983) (in Russian)
17. Drakunov, S.V.: Sliding-mode observers based on equivalent control method. In: Confrence on Decision and Control, pp. 2368–2369 (1992)
18. Barbot, J.P., Boukhobza, T., Djemai, M.: Sliding mode observer for triangular input form. In: Confrence on Decision and Control (1996)
19. Shtessel, Y., Edwards, C., Fridman, L., Levant, A.: Sliding Mode Control and Observation. Control Engineering Series. Birkhauser, NY (2014)
20. Utkin, V.I.: Sliding Modes in Problems of Optimization and Control, Nauka, Moscow (1981) (in Russian)
21. Levant, A.: Higher-order sliding modes, differentiation and output-feedback control. Int. J. Control **76**(9), 924–941 (2003)

Chapter 6
Chattering Problem

Abstract Due to different imperfections of switching devices, unmodeled dynamics, and discrete-time implementation, the state trajectories are not confined to the sliding manifold but run in its vicinity. Usually, it results in high-frequency oscillation with a finite amplitude. This motion is referred to as chattering. The motion in some vicinity of the sliding manifold with no high-frequency component can be caused by continuous approximation of discontinuous control. When implementing a continuous controller by power converters, a technique like pulse-width modulation (PWM) is considered to be exploited to adapt the control law to the discontinuous system inputs.

Keywords Chattering · Power converter

6.1 What is Chattering? Basic Ideas

Theoretical methods of analysis and design of sliding mode control constitute the scope of the book, while the numerous applications proved to be beyond consideration. We discuss the main obstacle for application only. Due to different imperfections of switching devices, unmodeled dynamics, and discrete-time implementation, the state trajectories are not confined to the sliding manifold but run in its vicinity. Usually, it results in high-frequency oscillation with a finite amplitude. This motion is referred to as *chattering*. The motion in some vicinity of the sliding manifold with no high-frequency component can be caused by continuous approximation of discontinuous control. In all cases, the size of the vicinity determines accuracy of the methods based on ideal models. Chattering is the main source of serious criticism of sliding mode control. Sometimes too serious. Chattering is the common and well-known phenomenon for several decades in the area of power converters, called *ripple*. The experts understood that ripple or chattering cannot be avoided in dynamic systems with discontinuous functions in motion equations and without dramatizing their effect developed efficient methods for reducing it. In the framework of sliding mode control theory, these methods were inherited and generalized along with the

© The Author(s), under exclusive license to Springer Nature Switzerland AG 2020   73
V. Utkin et al., *Road Map for Sliding Mode Control Design*,
SpringerBriefs in Mathematics,
https://doi.org/10.1007/978-3-030-41709-3_6

development of new ones. Again following the selected style of the book, the basic
ideas of the methods are outlined in this section.

It is known that the systems with a continuous control input and feedback gain,
tending to infinity, sooner or later become unstable at the presence of unmodeled
dynamics. Similar situation takes place in the systems with discontinuous control:
the local gain in the vicinity of discontinuity point tends to infinity, which leads to
instability; in the course of motion from the discontinuity point, the local gain is
decreasing and the diverging process terminates. This picture explains the origin of
chattering—high-frequency oscillations in the vicinity of a switching surface.

In the first attempts [1] to suppress chattering, the authors offered to replace
discontinuous function

$$u(s) = \begin{cases} u^+ & \text{if } s > 0 \\ u^- & \text{if } s < 0 \end{cases}$$

for $|s| \leq \Delta$ by continuous one

$$u(s) = u^- + \frac{u^+ - u^-}{2\Delta}(s + \Delta).$$

The new function tends to the original one with Δ tending to zero. However, decreasing Δ leads to the high gain $\frac{u^+ - u^-}{2\Delta}$, the unmodeled dynamics will be excited, and
chattering will appear. The wide linear zone $|s| \leq \Delta$ can eliminate high-frequency
oscillations, but low gain $\frac{u^+ - u^-}{2\Delta}$ results in a low accuracy.

As follows from the above picture of chattering origination, at the intuitive level
we can state that chattering amplitude is increasing with the range of discontinuity
(see also the analysis in [2, Chap. 8]). Then keeping difference

$$|u^+(x) - u^-(x)|$$

at minimal level, preserving sliding mode leads to decreasing the chattering amplitude. For this purpose, the state functions serving as upper and low estimates of functions in right-hand sides of motion equations are needed [3, Chap. 3]. For instance,
the magnitude of discontinuous control should be proportional to the norm of the state
in linear systems. It is interesting that control was a piecewise linear state function in
the first publications on variable structure control [4], and the chattering amplitude
was decreasing in the course of approaching the origin in the state space.

A similar idea of decreasing the amplitude of discontinuous control can be implemented in systems with high degree of uncertainty. As was shown in [2, Chap. 2],
the equivalent control is an average value of a real control with filtered out high-
frequency component, and it can be obtained with the help of a low-pass filter. Since

$$u^- < u_{eq} < u^+,$$

the extreme values of control can be reduced without violation of these inequalities.

Assume that in the system

$$\dot{x} = f(t, x) + b(t, x)u$$

with the scalar control input

$$u = -M \ \text{sign}(s(x)),$$

the condition $\nabla s(x)b = 1$ holds. Then

$$\dot{s} = F(t, x) - M \ \text{sign}(s), \quad |F(t, x)| \le F_0 = \text{const}$$

and the sliding mode exists on the surface $s = 0$, if $M > F_0$. The amplitude of chattering is determined by the parameter M. If the function $F(t, x)$ is known *a priori* and far from the extreme value, the magnitude of discontinuity M can be decreased and, as a result, the chattering amplitude can be decreased as well. Equivalent control

$$[\text{sign}(s)]_{eq} = F/M$$

is the solution to

$$\dot{s} = 0.$$

It can be obtained by a low-pass filter

$$\mu \dot{z} + z = \text{sign}(s), 0 < \mu << 1, z \approx (\text{sign}(s))_{eq}.$$

Since $|z|$ is decreasing function of $|F(t, x)|$, the chattering amplitude can be decreased if the magnitude of discontinuity is selected as a function of $|z|$, for example,

$$u = -(M_0 |z| + \delta)\text{sign}(s).$$

If the sliding mode does not exist, then

$$\text{sign}(s) = \pm 1, \quad |z| = 1$$

and control takes one of the extreme values needed for enforcing the sliding mode. After the sliding mode occurs, the control magnitude is decreasing depending on the level of $F(t, x)$ along with the chattering amplitude.

Assume that the condition

$$|F(t, x)| < M_0, M_0 = \text{const}$$

holds for the system with a scalar control u and state-dependent switching function s

$$\left.\begin{aligned} \dot{s}(t) &= f(t, x(t)) + u(t), \\ u(t) &= -(M_0|z(t)| + \delta)\, \text{sign}(s(t)), \end{aligned}\right\}$$

where $\delta > 0$ and the control u depends on the equivalent value of $\text{sign}(s)$, obtained by a low-pass filter

$$\mu\dot{z}(t) + z(t) = \text{sign}(s(t, x(t))), \quad \mu \ll 1, \quad z \approx [\text{sign}(s)]_{eq} = \frac{f}{M_0|z| + \delta}.$$

If $f > 0$, then $z > 0$ and

$$z^2 + \frac{\delta}{M_0} z - \frac{f}{M_0} = 0.$$

It follows from

$$z = -\frac{\delta}{2M_0} + \sqrt{\frac{\delta^2}{4M_0^2} + \frac{f}{M_0}} \leq -\frac{\delta}{2M_0} + \sqrt{\frac{\delta^2}{4M_0^2} + 1}$$

that $z < 1$. Similarly, it can be shown that $z > -1$ for $z < 0$. The condition

$$u^- < u_{eq} < u^+$$

is the condition for sliding mode to exist. It is the case for our system: $\left|[\text{sign}(s)]_{eq}\right|$ is between the extreme values $+1$ and -1. This value tends to zero with $f(t, x)$ tending to zero and as a result the magnitude of the discontinuous control $(M_0|z| + \delta)$ is decreased along with chattering amplitude.

Similar approach based on using equivalent control is considered in the next chapter, titled "Adaptive SMC", to minimize the discontinuity magnitude and decrease chattering.

The above two methodologies, based on replacing discontinuous control by continuous one and varying amplitude of discontinuous control in the boundary layer of switching surface, are not applicable for widely used power converters with outputs taking one of the two extreme values only.

An asymptotic observer in the control loop can eliminate chattering despite discontinuous control laws. The key idea as proposed in [5] is to generate ideal sliding mode in an auxiliary observer loop rather than in the main control loop. Ideal sliding mode is possible in the observer loop since the control is designed as a discontinuous function of the observer state, and it is entirely generated in the control software which does not contain any unmodeled dynamics. The main loop follows the observer loop according to the observer dynamics. Despite applying a discontinuous control signal with switching as the plant input, no chattering occurs and the system behaves as if an equivalent continuous control was applied. A block diagram for the system with disregarded actuator dynamics in the ideal mode ($\mu^2 \ll 1$) is shown in Fig. 6.1. The observer serves as a bypass for high-frequency component and the unmodeled dynamics are not excited.

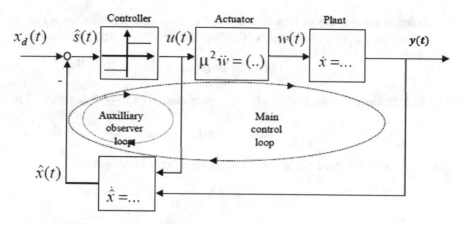

Fig. 6.1 Simplest model with fast actuator

For the simplest example

$$\dot{x}(t) = u(t), \quad u = -M \, \text{sign}(x)$$

the state x is reduced to zero in sliding mode. The state is known, but if it is corrupted by noise, it is reasonable to use an observer for filtration:

$$\frac{d}{dt}\hat{x}(t) = u(t) + l(x(t) - \hat{x}(t)),$$

where $l > 0$ is constant and $\hat{x}$ is the estimate of x. If

$$1/(\mu s + 1)^2, \quad 0 < \mu \ll 1$$

is a transfer function of the unmodeled dynamics of the actuator with output $w = w_1$, then

$$\left. \begin{aligned} \dot{x}(t) &= w_1(t), \\ \mu \dot{w}_1(t) &= w_2(t), \\ \mu \dot{w}_2(t) &= -2w_2(t) - w_1(t) + u(t), \\ u(t) &= -M \, \text{sign}(\hat{x}(t)). \end{aligned} \right\}$$

Since $\hat{x}$ and $\dfrac{d}{dt}\hat{x}(t)$ have opposite signs for a sufficiently high value M, sliding mode occurs:

$$\hat{x} \equiv 0, \quad u_{eq} = -lx$$

and we have

$$\left. \begin{aligned} \dot{x}(t) &= w_1(t), \\ \mu \dot{w}_1(t) &= w_2(t), \\ \mu w_2(t) &= -2w_2(t) - w_1(t) - lx. \end{aligned} \right\}$$

According to the singular perturbation theory [6], the solution of this linear system consists of fast and slow components. The first one decays rapidly such that after a short time

$$|w_1 + lx| = O(\mu)$$

with no high-frequency oscillation while the slow motion (if $\mu \to 0$) is governed by

$$\dot{x}(t) = -lx(t)$$

and decays with the same rate as the observer state in the ideal system

$$\bar{x} = \hat{x} - x, \quad \frac{d}{dt}\bar{x} = -l\bar{x}.$$

The control in the systems with observers is a discontinuous time function. It can lead to high wear or even destroying actuators with moving mechanical parts (like valves in pneumatic or hydraulic actuators). The destroying effect can be reduced considerably by inserting a low-pass filter to make an actuator input smooth, as it was demonstrated in Hsu [7] for model reference adaptive systems. The filter can be handled as unmodeled dynamics, and it does not affect the ideal sliding mode in the model loop. As a result, the system is free of chattering. It is important to note that the output of the filter is close to the equivalent control with mismatch depending on the filter time constant and the discontinuous control amplitude [3, Chap. 2]. If the equivalent control can be estimated, then the amplitude of the discontinuous control can be decreased (recall that trajectories belong to the discontinuity surface for control $u = u_{eq}$). Therefore, the effect of unmodeled dynamics in the plant model can be reduced considerably.

Demonstrate the chattering suppression effect in systems with an observer for the second-order system

$$\ddot{x} = -\dot{x} - x + w$$

with disregarded dynamics of the actuator

$$\left.\begin{array}{c} w(\mathrm{p}) = \dfrac{1}{(\mu\mathrm{p} + 1)^2}u(\mathrm{p}), \ 0 < \mu << 1, \\ \mu^2\ddot{w} = -2w\dot{w} - w + u, \\ u = -u_0\mathrm{sign}(s), \ s = cx + \dot{x}, \\ u_0 \text{ and } c \text{ are positive constants}, \mathrm{p} \text{ is Laplace variable.} \end{array}\right\}$$

The system with the observer practically is free of chattering and similar to the ideal system with $\mu = 0$ (Fig. 6.2).

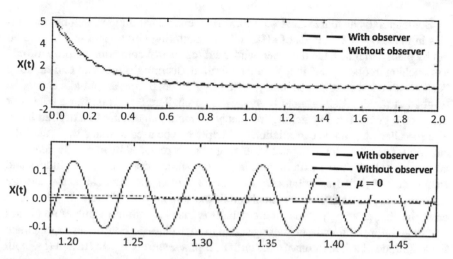

Fig. 6.2 Chattering with unmodeled dynamics (upper plots) and zoomed processes with and without unmodeled dynamics

6.2 Power Converters

The main disadvantage of the observer-based chattering suppression method is need of the plant equations. The influence of uncertainties can be reduced by increasing the observer gains, which is not dangerous from the point of stability, since an observer is implemented in software part of control and free of unmodeled dynamics. The natural way of reducing the chattering effect is increasing the oscillation frequency to be filtered out by plant dynamics. When implementing a continuous controller by power converters, a technique like pulse-width modulation (PWM) has to be exploited to adapt the control law to the discontinuous system inputs. In the light of recent advances of high-speed circuitry and insufficient linear control methodologies for internally nonlinear high-order plants such as AC motors, sliding mode control has become increasingly popular, since this control methodology implies discontinuous control actions and on/off is the only admissible mode for *power converters*. Therefore, outputs of power converters can be used as control action directly without PWM. Due to the switching element (Fig. 6.3), the control is discontinuous and can be selected as a function of the mismatch between real and desired values of the output voltage. The tracking problem is solved if sliding mode is enforced.

Fig. 6.3 Power converter

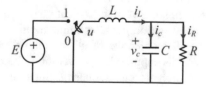

Commercially available electronic converters enable handling switching frequencies in the range of hundreds of kHz, and the chattering component will be filtered out by plant dynamics. On the other hand, heat losses in a converter are proportional to switching frequency and they should be limited. Seemingly, the dead-end situation is encountered: the desired accuracy, depending on chattering amplitude, cannot be reached at maximal admissible frequency. The challenge is to design a converter based on a given, fixed switching frequency, reducing chattering to a desired level. The so-called "harmonic cancellation" principle can be used to find a way out of the situation. The converter with the desired output u_0 is replaced by m parallel identical converters called phases with outputs u_0/m and summation block. Amplitudes and frequencies of periodic oscillations with period T or chattering in all converters are the same, but phases depend on initial conditions and may be different. For the worst case, the chattering amplitude after summation takes maximum value, if all phases are equal. Representing the chattering signal in each phase as Fourier series, calculate the amplitude of kth harmonic in the output, if phase shift between ith and $(i + 1)$th phases is equal to T/m:

$$\omega = \frac{2\pi}{T}, \omega_k = k\omega,$$

$$\left.\begin{array}{c} \sum_{i=0}^{k-1} \sin\left[\omega_k\left(t - \frac{2\pi}{\omega m}i\right)\right] = \sum_{i=0}^{m-1} Im\left[e^{j\left(\omega_k t - \frac{2\pi k}{m}i\right)}\right] \\ = Im\left[e^{j\omega_k t}Z\right], \quad Z = \sum_{i=0}^{m-1} e^{-j\frac{2\pi k}{m}i}. \end{array}\right\}$$

Function Z satisfies condition $Z = Ze^{-j\frac{2\pi k}{m}}$ and is different from zero if $e^{-j\frac{2\pi k}{m}} = 1$ or for $k = m, 2m, 3m, \ldots\ldots$ It means that at least the first $m - 1$ harmonics are canceled and increasing number of phases will decrease the chattering amplitude essentially.

To get this effect, called harmonic cancellation, the switching commands should be selected such that phase shifts between two neighbor phases are equal to T/m. Switching elements of power converters are implemented with state-dependent width of hysteresis loops and the same magnitudes. For the first phase, it is selected such that the switching frequency is equal to maximal admissible value. If the switching function of the second phase depends not only on the tracking error (differences between real and desirable output values) but also on the error of the first phase, then chattering frequency of the second phase coincides with that of the first phase. The width of hysteresis loop in the second phase is selected such that the phase shift is equal to T/m. The subsequent phases are designed in similar ways. Theoretical substantiation and experimental results may be found in [3, Chap. 8], [8]. The results of simulation of the four-phase power converter with parameters in Table 6.1 are shown in Fig. 6.4.

In spite of high level of chattering in each phase, it is practically deleted in the converter output after summation.

The objective of all the above methods is estimation of chattering amplitude and its suppression. In addition, the harmonic cancellation method implies control of a chattering frequency. Since as a rule chattering is far from being harmonic

Table 6.1 Parameters of power converter

Parameter	Value
$L(H)$	1
$C(F)$	1
$R_a(\Omega)$	1
$R_L(\Omega)$	1
$V_s(V)$	12

Fig. 6.4 Four-phase converter, $u_{0ref} = 4.2$ V

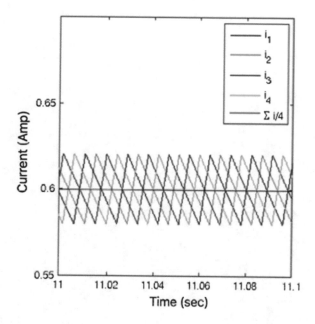

oscillations, it is difficult to find the solution of the corresponding differential equation as a time function. Frequency analysis proved to be an efficient approximate tool to solve the problem under two main assumptions: chattering is a periodic time function and high-order harmonics can be disregarded. In the framework of this approach, *describing function method* and its further development can be found in [9].

References

1. Slotine, J.J., Li, W.: Applied Nonlinear Control. Prentice Hall (1991)
2. Utkin, V.: Sliding Modes in Control and Optimization. Springer, Berlin (1992)
3. Utkin, V., Guldner, J., Shi, J.: Sliding Mode Control in Electro-Mechanical Systems. CRC Press (2009)
4. Emelyanov, S. (ed.): Theory of Variable Structure Control Systems. Nauka (1970) (in Russian)
5. Bondarev, A., Bondarev, S., Kostyleva, N., Utkin, V.: Sliding mode in systems with asymptotic state observers. Autom. Remote Control **46**, 49–64 (1985)

6. Kokotovic, P., Khalil, H., O'Relly, J.: Singular Perturbation Method in Control: Analysis and Design. Academic Press (1986)
7. Hsu, L.: Smooth sliding mode control of uncertain systems based on prediction error. Int. J. Robust Nonlinear Control 7(4), 353–372 (1997)
8. Biel, D., Fossas, E.: Some experiments on chattering suppression in power converters. In: 18th IEEE International Conference on Control Applications, pp. 1523–1529 (2009)
9. Boiko, I.: Discontinuous Control Systems : Frequency-domain Analysis and Design. Birkhauser, Boston (2009)

Chapter 7
High-Order Sliding Mode Control

Abstract The high-order sliding mode (HOSM) is discussed here as an alternative methodology to the conventional sliding mode control to design feedback control, observers, and to handle the chattering problem. Here, the principle ideas of HOSM methodology are presented. Similar to the second-order sliding mode, HOSM implies that sliding mode occurs in the manifold of a smaller order after a finite time interval.

Keywords High-order sliding mode · Finite-time convergence

7.1 HOSM Definition and Its Properties

The core idea of HOSM control design implies existence of sliding mode in the manifold on $n - r$ order ($1 < r < n - 1$) after a finite time interval similar to systems in Sect. 3.9 with sliding mode of the second order.

The next definition of HOSM is taken from [1].

Definition 7.1 Let σ be an output of the closed-loop system (1), and there exists $r \in \mathbb{N}$ such that

- for $p = 1, 2, \ldots, r - 1$ total time derivatives $\sigma^{(p)}$ along the trajectories of the closed-loop system (2.1) are continuous functions of time and state variables;
- the set
$$S = \left\{ \sigma^{(p)} = 0 \text{ for any } p = 0, 1, 2, \ldots, r - 1 \right\}$$

  is the sliding manifold of the closed-loop system (2.1);
- $\sigma^{(r)}$ is discontinuous in a vicinity of S.

Then S is the rth-order sliding manifold.

The efficiency of SMC has been demonstrated for a set of applications [2, Chap. 8], [3, Chaps. 10–13].

From the application standpoint, as stated in [4], it is important for the conventional sliding mode that

© The Author(s), under exclusive license to Springer Nature Switzerland AG 2020   83
V. Utkin et al., *Road Map for Sliding Mode Control Design*,
SpringerBriefs in Mathematics,
https://doi.org/10.1007/978-3-030-41709-3_7

– the implementation of SM implies that the relative degree (RD) of the system should be equal to one which is considered as a serious constraint for SM applications;
– the discontinuities in control lead to chattering—an unacceptable phenomenon in some applications.

High-order sliding mode (HOSM) was offered as an alternative methodology to the conventional sliding mode control. According to [4, Chap. 8], "the intrinsic difficulties of conventional SMC are mitigated by HOSM controllers and this approach is effective for arbitrary relative degree, and the well-known chattering is considerably reduced". The principle ideas of HOSM methodology and references can be found also in [4, Chap. 8]. Similar to the second-order sliding mode, HOSM implies that sliding mode occurs in the manifold of the $(n - k)$th order $(1 < k < n)$ after a finite time interval.

7.2 Simple Examples of Systems with HOSM Control

First, consider the system (3.12) with sliding mode control of the second order. Let the system output σ be governed by equation

$$\ddot{\sigma} = u + F, \tag{7.1}$$

where F is a bounded function of the state and time.

As shown in Sect. 3.9, the state can be reduced to the origin in the plane $(\sigma, \dot{\sigma})$ after a finite time interval by twisting and super-twisting algorithms and the second-order sliding mode arises. The authors of HOSM control offered two more versions to enforce sliding modes of the second order:

$$\left. \begin{array}{c} u = -M \mathrm{sign}(s), \quad M > 0, \\[2mm] s := \dot{\sigma} + \alpha \sqrt{|\sigma|}\, \mathrm{sign}\,(\sigma), \quad \alpha > 0, \end{array} \right\} \tag{7.2}$$

or

$$u = -\rho \frac{s}{|\dot{\sigma}| + \sqrt{|\sigma|}}. \tag{7.3}$$

For the *first version* (7.2), the origin is reached in sliding mode after finite time. The state plane is similar to that in Fig. 4.2. For the *second version* (7.3), the origin is reached after finite time [4, Chap. 4] but without intermediate sliding mode (Fig. 7.1).

Both versions were generalized for the systems with sliding modes of an arbitrary order r and finite time of reaching the sliding manifold.

Here are control algorithms for the third-order sliding modes:

$$u = -\alpha \mathrm{sign}(\ddot{\sigma} + 2\left(|\dot{\sigma}|^3 + |\sigma|^2\right)^{1/6} \mathrm{sign}\left(\dot{\sigma} + |\sigma|^{2/3}\, \mathrm{sign}\,(\sigma)\right)) \tag{7.4}$$

Fig. 7.1 HOSM without sliding mode in the reaching interval

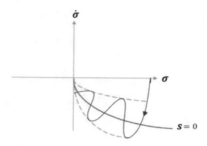

and

$$u = \alpha \frac{\ddot{\sigma} + 2\left(|\dot{\sigma}| + |\sigma|^{2/3}\right)^{-1/2} \left(\dot{\sigma} + |\sigma|^{2/3} \operatorname{sign}(\sigma)\right)}{|\ddot{\sigma}| + 2\sqrt{\left(|\dot{\sigma}| + |\sigma|^{2/3}\right)}}. \tag{7.5}$$

It is interesting that the discontinuous controls similar to (7.3) and (7.5) have one discontinuity point $(\sigma, \dot{\sigma}, \ldots, \sigma^{(r-1)})$ only. Implementation of all HOSM controls (except for super-twisting) needs all time derivative $(\dot{\sigma}, \ldots, \sigma^{(r-1)})$ of the output σ.

7.3 HOSM in SISO Affine Systems in Regulation and Tracking Problems

7.3.1 Uncertain Plant

Following [5], consider the nonlinear system

$$\left. \begin{aligned} \dot{z} &= f(t, z) + g(t, z)u, \\ y &= h(t, z), \end{aligned} \right\} \tag{7.6}$$

where $z \in R^n$ defines the state vector, $u \in R$ is the control input, $y \in R$ is the output, and $h(t, z) : R \times R^n \to R$ is a smooth output function. The functions $f(t, z)$ and $g(t, z)$ are uncertain smooth vector fields on R^n.

A standard problem of control is the output tracking problem, consisting in forcing the output y to track a (time-varying) signal $y_R(t)$, namely, in making the output $\sigma = y - y_R$ vanishing in finite time and to keep $\sigma \equiv 0$ exactly by a bounded (discontinuous) feedback control. All differential equations are understood in the Filippov's sense.

When the relative degree ρ with respect to σ is known, well defined and constant this is equivalent to designing a controller for the differential inclusion (*DI*)

$$\left.\begin{array}{l} \dot{x}_i = x_{i+1}, i = 1, \dots, \rho - 1 \\[2mm] \dot{x}_\rho \in [-C, C] + [K_m, K_M]u \end{array}\right\} \sum_{DI} \qquad (7.7)$$

where

$$x = (x_1, \dots, x_\rho)^\mathsf{T} = (\sigma, \dot{\sigma}, \dots, \sigma^{(\rho-1)})^\mathsf{T},$$

$$\sigma^{(i)} = \frac{d^i}{dt^i} h(z, t) - \frac{d^i}{dt^i} y_R$$

(if such $\sigma^{(i)}$ ($i = 1, \dots, \rho - 1$) exist). Below based on smooth control and Lyapunov functions, we derive a family of homogeneous HOSM controllers.

7.3.2 Lyapunov Function for HOSM Controllers' Analysis

Given the relative degree $\rho \geq 2$, assign the homogeneity weights $r_i = \rho - i + 1$ to the variables x_i, obtaining the dilatation vector $\mathbf{r} = (\rho, \rho - 1, \dots, 1)$.[1] Define also an arbitrary non-decreasing sequence of positive real numbers α_i, so that

$$0 \leq \alpha_1 \leq \cdots \leq \alpha_{\rho-1} \leq \alpha_\rho .$$

Furthermore, we define recursively, for $i = 2, \dots, \rho$, the r-homogeneous functions ($k_i > 0$) from C^1:

$$\left.\begin{array}{l} \sigma_1(\bar{x}_1) = \lceil x_1 \rfloor^{\frac{\rho+\alpha_1}{\rho}}, \dots, \ \sigma_i(\bar{x}_i) = \lceil x_i \rfloor^{\frac{\rho+\alpha_i}{\rho-i+1}} + k_{i-1}^{\frac{\rho+\alpha_i}{\rho-i+1}} \lceil \sigma_{i-1} x_{i-1} \rfloor^{\frac{\rho+\alpha_i}{\rho+\alpha_{i+1}}}, \\[3mm] \bar{x}_i = (x_1, \dots, x_i)^\mathsf{T}, \ \lceil x_i \rfloor^s := |x_i|^s \operatorname{sign}(x_i) . \end{array}\right\}$$

For any constant $m \geq \max_{1 \leq i \leq \rho}\{2\rho + 1 + \alpha_{i-1} - i\}$ we also define recursively, for $i = 2, \dots, \rho$, the C^1 r-homogeneous functions

$$\left.\begin{array}{l} V_1(x_1) = \frac{\rho}{m}|x_1|^{\frac{m}{\rho}}, \dots, \ V_i(\bar{x}_i) = \gamma_{i-1} V_{i-1}(\bar{x}_{i-1}) + W_i(\bar{x}_i) \\[3mm] W_i(\bar{x}_i) = \frac{r_i}{m}|x_i|^{\frac{m}{r_i}} - \lceil \nu_{i-1}(\bar{x}_{i-1}) \rfloor^{\frac{m-r_i}{r_i}} x_i + \left(1 - \frac{r_i}{m}\right)|\nu_{i-1}(\bar{x}_{i-1})|^{\frac{m}{r_i}} \\[3mm] \nu_1(x_1) = -k_1 \lceil \sigma_1 \rfloor^{\frac{r_2}{\rho+\alpha_1}} = -k_1 \lceil x_1 \rfloor^{\frac{\rho-1}{\rho}}, \dots, \ \nu_i(\bar{x}_i) = -k_i \lceil \sigma_i(\bar{x}_i) \rfloor^{\frac{r_i+1}{\rho+\alpha_i}} \end{array}\right\}$$

---

[1] Definitions of homogeneity and dilation are given in Chap. 4 (see Definitions 4.1 and 4.2).

with (arbitrary) constants $\gamma_i > 0$. Here, $\sigma_i(\bar{x}_i)$, $\nu_i(\bar{x}_i)$, and $V_i(\bar{x}_i)$ are r-homogeneous of degrees $\rho + \alpha_i$, $r_i + 1$ and m, respectively. So, $V_c(x) = V_\rho(\bar{x}_\rho)$ is a smooth Lyapunov function for the considered uncertain plant (7.6).

7.3.3 HOSM Controllers

From $V_c(x)$, we can derive different controllers. In particular, we obtain the following family of discontinuous $(u_{\rho D})$ and quasi-continuous $(u_{\rho Q})$ controllers

$$\left. \begin{array}{l} u_{\rho D} = -k_\rho \varphi_D(x) = -k_\rho \left\lceil \sigma_\mu(x) \right\rfloor^0 \\[4mm] u_{\rho Q} = -k_\rho \varphi_Q(x) = -k_\rho \dfrac{\sigma_\rho(x)}{M(x)}, \end{array} \right\}$$

where $M(x)$ is any continuous r-homogeneous positive-definite function of degree $\rho + \alpha_\rho$, and (for simplicity) we assume that it is scaled so that $\left| \dfrac{\sigma_\rho(x)}{M(x)} \right| \leq 1$. The homogeneous controllers are derived by imposing the condition

$$\frac{\partial V_c(x)}{\partial x_\rho} \varphi(x) > 0$$

at all points where $\dfrac{\partial V_c(x)}{\partial x_\rho} \neq 0$, resulting from the condition $\dot{V}_c(x) < 0$ and in view of the relation

$$\dot{V}_c(x) = \frac{\partial V_c(x)}{\partial x_\rho} \dot{x}_\rho \in \frac{\partial V_c(x)}{\partial x_\rho} \left([-C, C] + [K_m, K_M] u_\rho\right)$$

$$\leq 2C \left| \frac{\partial V_c(x)}{\partial x_\rho} \right| - k_\rho K_m \frac{\partial V_c(x)}{\partial x_\rho} \varphi(x) \implies \frac{\partial V_c(x)}{\partial x_\rho} \varphi(x) > 0.$$

The values of k_i, for $i = 1, \ldots, \rho - 1$ can be fixed depending only on ρ and α_i.

7.3.4 Discontinuous and Quasi-Continuous HOSM Controllers for SISO Systems

Varying the selection of the free parameters $0 \leq \alpha_1 \leq \cdots \leq \alpha_\rho$, we obtain different families of controllers (for the orders of the relative degrees $\rho = 2, 3$).

Discontinuous Controllers

– *Nested Sliding Controllers* (when some of the α_i are different)

$$u_{2D}^{Nest} = -k_2 \left\lceil \lceil x_2 \rfloor^{2+\alpha_2} + k_1^{2+\alpha_2} \lceil x_1 \rfloor^{\frac{2+\alpha_2}{2}} \right\rfloor^0,$$

$$u_{3D}^{Nest} = -k_3 \left\lceil \lceil x_3 \rfloor^{3+\alpha_3} + k_2^{3+\alpha_3} \left\lceil \lceil x_2 \rfloor^{\frac{3+\alpha_2}{2}} + k_1^{\frac{3+\alpha_2}{2}} \lceil x_1 \rfloor^{\frac{3+\alpha_2}{3}} \right\rfloor^{\frac{3+\alpha_3}{3+\alpha_2}} \right\rfloor^0.$$

– *Relay Polynomial Controllers* (when $\alpha_\rho = \alpha_{\rho-1} = \cdots = \alpha_1 = \alpha \geq 0$)

$$u_{2D}^{Rel} = -k_2 \text{sign} \left(\lceil x_2 \rfloor^{2+\alpha} + \bar{k}_1 \lceil x_1 \rfloor^{\frac{2+\alpha}{2}} \right),$$

$$u_{3D}^{Rel} = -k_3 \text{sign} \left(\lceil x_3 \rfloor^{3+\alpha} + \bar{k}_2 \lceil x_2 \rfloor^{\frac{3+\alpha}{2}} + k_1 \lceil x_1 \rfloor^{\frac{3+\alpha}{3}} \right).$$

Quasi-Continuous Controllers

– *Nested Sliding Controllers* (when some of the α_i are different, $\beta_i > 0$)

$$u_{2Q}^{Nest} = -k_2 \frac{\lceil x_2 \rfloor^{2+\alpha_2} + k_1^{2+\alpha_2} \lceil x_1 \rfloor^{\frac{2+\alpha_2}{2}}}{|x_2|^{2+\alpha_2} + \beta_1 |x_1|^{\frac{2+\alpha_2}{2}}},$$

$$u_{3Q}^{Nest} = -k_3 \frac{\lceil x_3 \rfloor^{3+\alpha_3} + k_2^{3+\alpha_3} \left\lceil \lceil x_2 \rfloor^{\frac{3+\alpha_2}{2}} + k_1^{\frac{3+\alpha_2}{2}} \lceil x_1 \rfloor^{\frac{3+\alpha_2}{3}} \right\rfloor^{\frac{3+\alpha_3}{3+\alpha_2}}}{|x_3|^{3+\alpha_3} + \beta_2 |x_2|^{\frac{3+\alpha_3}{2}} + \beta_1 |x_1|^{\frac{3+\alpha_3}{3}}}.$$

– *Relay Polynomial Controllers* (when $\alpha_i = \alpha \geq 0$)

$$u_{2Q}^{Rel} = -k_2 \frac{\lceil x_2 \rfloor^{2+\alpha} + k_1^{2+\alpha} \lceil x_1 \rfloor^{\frac{2+\alpha}{2}}}{|x_2|^{2+\alpha} + \beta_1 |x_1|^{\frac{2+\alpha}{2}}},$$

$$u_{3Q}^{Rel} = -k_3 \frac{\lceil x_3 \rfloor^{3+\alpha} + \bar{k}_2 \lceil x_2 \rfloor^{\frac{3+\alpha}{2}} + \bar{k}_1 \lceil x_1 \rfloor^{\frac{3+\alpha}{3}}}{|x_3|^{3+\alpha} + \beta_2 |x_2|^{\frac{3+\alpha}{2}} + \beta_1 |x_1|^{\frac{3+\alpha}{3}}}.$$

All these controllers solve the problem posed before.

Theorem 7.1 ([5]) *For any relative degree $\rho \geq 2$ between x_1 and control, each controller of the families of discontinuous or quasi-continuous controllers with arbitrary parameters $0 \leq \alpha_1 \leq \cdots \leq \alpha_\rho$ is ρ-sliding homogeneous, and for k_ρ sufficiently large the ρth-order sliding mode*

$$x = (x_1, \ldots, x_\rho)^{\mathsf{T}} = (\sigma, \dot{\sigma}, \ldots, \sigma^{(\rho-1)})^{\mathsf{T}} = 0$$

is established in finite time for the uncertain system (7.6)–(7.7) under properly chosen gains $k_1, \ldots, k_{\rho-1}$ and $\beta_1, \ldots, \beta_{\rho-1}$.

Conclusion

In the framework of HOSM study, several interesting research directions were initiated by A. Levant (basic concepts, feedback design, observers [1, 6]), J. Moreno (stability of systems with HOSM [7, 8]), and I. Boiko (frequency methods [9]). HOSM gave born to many mathematical problems. All authors of this book published solutions to some of them (see, for example, [10–14]). The interested reader can find the arguments confirming the stated advantages of HOSM control in [4, Chap. 8].

References

1. Levant, A.: Higher-order sliding modes, differentiation and output-feedback control. Int. J. Control **76**(9), 924–941 (2003)
2. Edwards, C., Spurgeon, S.: Sliding Mode Control: Theory and Applications. Taylor and Francis (1998)
3. Utkin, V., Guldner, J., Shi, J.: Sliding Mode Control in Electro-Mechanical Systems. CRC Press (2009)
4. Shtessel, Y., Edwards, C., Fridman, L., Levant, A.: Sliding Mode Control and Observation. Control Engineering Series. Birkhauser, NY (2014)
5. Cruz-Zavala, E., Moreno, J.A.: Homogeneous high order sliding mode design: a lyapunov approach. Automatica **80**, 232–238 (2017)
6. Levant, A.: Sliding order and sliding accuracy in sliding mode control. Int. J. Control **58**(6), 1247–1263 (1993)
7. Moreno, J., Osorio, M.: A lyapunov approach to second-order sliding mode controllers and observers. In: 47th IEEE Conference on Decision and Control (CDC), pp. 2856–2861 (2008)
8. Moreno, J.: A lyapunov approach to output feedback control using second-order sliding modes. IMA J. Math. Control Inf. **29**(3), 291–308 (2012)
9. Boiko, I.: Discontinuous Control Systems : Frequency-domain Analysis and Design. Birkhauser, Boston (2009)
10. Utkin, V.: Mechanical energy-based lyapunov function design for twisting and super-twisting sliding mode control. IMA J. Math. Control Inf. **32**(4), 675–688 (2015)
11. Utkin, V.I.: Discussion aspects of high-order sliding mode control. IEEE Trans. Autom. Control **61**(3), 829–833 (2016)
12. Polyakov, A., Poznyak, A.: a). Lyapunov function design for finite-time convergence analysis: "twisting" controller for second-order sliding mode realization. Automatica **45**, 444–448 (2009)
13. Orlov, Y.: Finite time stability and robust control synthesis of uncertain switched systems. SIAM J. Control Optim. **43**(4), 1253–1271 (2005)
14. Orlov, Y.: Discontinuous Systems: Lyapunov Analysis and Robust Synthesis Under Uncertainty Conditions. Springer, Berlin (2008)

Chapter 8
Discrete-Time Systems

Abstract Discretization of continuous-time models is considered. Definition of SM for discrete-time systems is presented. The behavior under uncertainties of discrete-time systems, controlled by SM, is briefly discussed.

Keywords Discretization of continuous-time models · The definition of SM for discrete-time systems · The behavior under uncertainties

8.1 Discretization of Continuous-Time Models

The design principles of sliding mode control were developed mainly for finite-dimensional continuous-time systems. At the same time, more and more systems are implemented based on digital microcontrollers. In both discrete- and continuous-time cases, the control is assumed to take one of the finite number of values. The ideal siding mode in a continuous-time system can be obtained with switching frequency tending to infinity. But it cannot be higher than a sampling frequency in discrete-time systems. As a result, the state oscillates in a vicinity of the switching surface and the oscillations are referred to as chattering as well, although they are of different nature than those caused by unmodeled dynamics. Natural way to increase the chattering frequency and decrease its amplitude is decreasing a sampling time, but it may prove to be too short for calculation of a control signal. Sliding mode with $x(t) \equiv 0$ occurs in the simple first-order system

$$\left. \begin{array}{l} \dot{x}(t) = f(t, x(t)) + u(t), \quad t > 0, \\ |f(t, x)| < f_0 \in R, \quad u(t) = -u_0 \text{sign}(x(t)), \quad f_0 < u_0 \in R \end{array} \right\} \quad (8.1)$$

after a finite time interval. Figure 8.1 illustrates chattering caused by discrete-time implementation of the control for explicit Euler approximation of the above system

$$x_{k+1} = x_k + \delta(f(k\delta, x_k) - u_0 \text{sign}(x_k)),$$

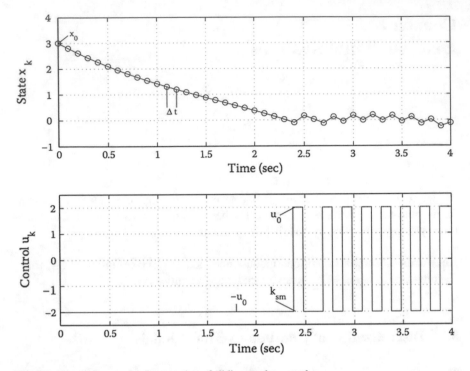

Fig. 8.1 Direct discrete implementation of sliding mode control

where δ is a sampling time. Since the control is constant within the sampling interval, chattering appears always starting from k_{sm}, if the control is a discontinuous state function (except for unlikely cases $f = u_0$ or $f = -u_0$), even if function f is known. This motion was called the "*zigzag regime*" in [1] (see Fig. 8.1).

Let us now discuss the methods, offered in the literature, which intend to reduce chattering amplitude for n-dimensional linear discrete-time systems with m-dimensional control

$$x_{k+1} = Ax_k + Bu_k + d_k, \tag{8.2}$$

where A and B are matrices with bounded elements, and d_k is a disturbance vector with bounded elements. Control is selected as a piecewise linear state function in [2], $u_k = Kx_k$, with elements of K undergoing discontinuities on planes

$$s_i(x) = c_i^T x = 0, \ c_i \in R^n, \ i = 1, \ldots, m.$$

If the disturbance is equal to zero and the motion in manifold

$$s(x) = (s_1(x), \ldots, s_m(x)) = 0$$

is stable with the state tending to zero, then the system can be stabilized by a proper choice of matrix K. The state reaches $s(x) = 0$, and its trajectories are converging to the origin with chattering amplitude tending to zero [2]. Of course the stabilization problem cannot be solved at the presence of disturbance d_k, since $x_k = 0$ cannot be a solution to (8.2).

The chattering amplitude (Fig. 8.1) for system (8.1) is increasing with increasing magnitude u_0 of the discontinuous control. For control as sum of linear and discontinuous functions of s, the magnitude of the second term can be decreased and as a result it decreases the chattering amplitude [3].

Replacing discontinuous function by some continuous state function in a boundary layer of manifold $s(x) = 0$ is also the way to reduce chattering caused by discretization of the control [4]. The control is selected such that $s(x_{k+1}) = s(x_k)$ in some vicinity of $s(x) = 0$. It is called the equivalent control u_{eq} similar to continuous-time systems with $\dot{s}(x(t)) = 0$ for $s(x(t)) = 0$ and $u(t) = u_{eq}(t)$. Having reached the vicinity, the state remains in it without chattering. Implementation of such selected control implies that the disturbance is known; otherwise, $s(x_{k+1})$ cannot be found. The upper estimate of the boundary layer with state trajectories of so-called quasi-sliding mode is found in [4] for systems with known bounds of disturbances only. The width of the boundary layer or the amplitude of chattering can be reduced if compared with those for direct implementation of discontinuous control.

As follows from the titles of cited above paper, analysis and design methods are studied for discrete-time systems with sliding modes. At the same time, the concept "discrete-time sliding mode" needs a strict definition as well as the conditions for this motion to exist are needed for developing the theory of this class of control systems.

8.2 Definition of SM for Discrete-Time Systems

For continuous-time systems, state trajectories belong to manifold $s(x) = 0$ and the manifold is reached after a finite time interval. These properties are formalized in Definition 2.1 for continuous-time systems. The definition in terms of discrete-time systems

$$x_{k+1} = F(x_k), \quad F : R^n \to R^n \tag{8.3}$$

is formulated in [5].

Definition 8.1 Discrete-time sliding mode takes place on subset Σ of manifold

$$s(x) = 0, \ s : R^n \to R^m, \ m < n,$$

if there exists an open neighborhood U of this subset such that $s(F(x)) \in \Sigma$ for any $x \in U$ (see Fig. 8.2).

Fig. 8.2 Discrete-time
sliding manifold

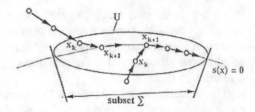

It makes sense to note that the idea to give a definition in terms of finite-time convergence works for both continuous- and discrete-time systems. In contrast to continuous-time systems, sliding mode may arise in discrete-time systems with continuous right-hand side in (8.3) satisfying Lipschitz condition even for a linear function $F(x) = Ax$, where A is a constant matrix. Sliding mode exists on linear manifold

$$s(x) = Cx = 0$$

in an affine system

$$x_{k+1} = F(k, x_k) + B(k, x_k)u_k, \quad x_k \in R^n, \quad u_k \in R^m, \tag{8.4}$$

if $s(x_{k+1}) = 0$ for any x_k. This condition holds for control

$$u_k = u_k^{eq} = -(CB(k, x_k))^{-1} CF(k, x_k) \tag{8.5}$$

assuming that matrix $CB(k, x_k)$ is non-singular. Control (8.5) is called equivalent control, but it is different from that selected in [4] to fulfill the condition $s(x_{k+1}) = s(x_k)$. Control (8.5) makes $s(x)$ equal to zero after one step [5–8]. It is not realistic to implement control (8.5) for small values of a sampling time δ, if Eq. (8.4) is an approximation of a continuous-time system

$$\dot{x}(t) = f(t, x(t)) + b(t, x(t))u(t), t > 0.$$

Then

$$F(k, x_k) = x_k + \delta f(k\delta, x_k),$$
$$B(k, x_k) = \delta b(k\delta, x_k),$$
$$s(x_{k+1}) = s(x_k) + \delta(Cf(k, x_k) + CB(k, x_k)u_k)$$

and

$$u_k^{eq} = -\frac{1}{\delta}(CB(k, x_k))^{-1} s(x_k) - (CB(k, x_k))^{-1} CF(k, x_k). \tag{8.6}$$

For any $s(x_k) \neq 0$, the control tends to infinity with δ tending to zero. It means that control (8.6) should be modified taking into account control constraints existing in

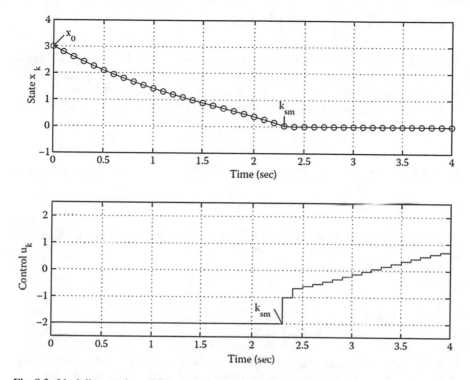

Fig. 8.3 Ideal discrete-time sliding mode control in system with known parameters

any real-life system. For control bounded by norm $\|u_k\| < \tilde{u} = \text{const}$, this constraint is taken into account in [6]

$$
u_k = \begin{cases} u_k^{eq} & \text{if } \|u_k^{eq}\| \le \tilde{u}, \\ \tilde{u} \dfrac{u_k^{eq}}{\|u_k^{eq}\|} & \text{if } \|u_k^{eq}\| > \tilde{u}. \end{cases} \tag{8.7}
$$

Control (8.7) is collinear to u^{eq}, and its norm does not exceed the admissible value. Control should be equal to the equivalent control on the manifold $s(x) = 0$; hence, the condition

$$
\left\| (CB(k, x_k))^{-1} CF(k, x_k) \right\| < \tilde{u} \tag{8.8}
$$

should hold. As follows from (8.6), the control is equal to u^{eq} in δ—vicinity of manifold $s(x) = 0$. As shown in Fig. 8.3, x_k for the discrete-time version of system (8.1) is equal to zero identically after a finite number of steps.

8.3   Behavior Under Uncertainties

Complete information on plant parameters and disturbance is needed for implementation of the equivalent control. Suppose now that the system operates under uncertainty conditions: matrices A and B and disturbance d in (8.2) may vary in some ranges. Suppose that the input matrix can be represented in the form

$$B = B_0 + \Delta B\,(k, x_k)$$

with known matrix B_0. Then the condition similar to (8.8) can be written in the form

$$\tilde{u}\left(1 - \left\|(CB_0)^{-1}\,C\Delta B\,(k, x_k)\right\|\right) > \left\|(CB(k, x_k))^{-1}\,CF(k, x_k)\right\|.$$

The range of admissible variations $\Delta B\,(k, x_k)$ of matrix B can be found from this inequality.

Similar to (8.7), control

$$u_k = \begin{cases} -(CB_0)^{-1}s(x_k) & \text{if } \left\|(CB_0)^{-1}s(x_k)\right\| \le \tilde{u}, \\[2mm] -\tilde{u}\,\dfrac{(CB_0)^{-1}s(x_k)}{\left\|(CB_0)^{-1}s(x_k)\right\|} & \text{if } \left\|(CB_0)^{-1}s(x_k)\right\| > \tilde{u} \end{cases} \tag{8.9}$$

is in the admissible domain but it depends on neither plant parameters nor disturbances. Again, $s(x_k)$ is monotonously decreasing and after finite number of steps control

$$u_k = -(CB_0)^{-1}s(x_k)$$

will belong to the admissible domain and the subsequent motion will be in δ-vicinity of manifold $s(x) = 0$, while the control values remain in the admissible domain. This result should be expected for the systems operating under uncertainty conditions, because control is constant within a sampling time, while the parameters and disturbance can take any value from some ranges. In contrast to discrete-time systems with discontinuous control, the systems with control (8.9) are free of chattering, because the control is a continuous state function for the motion within delta vicinity of $s = 0$.

References

1. Milosavlevic, C.: General conditions for the existence of a quasi-sliding mode on the switching hyperplane in discrete variable structure systems. Autom. Remote Control **46**(3), 307–314 (1985)
2. Sarpturk, S.Z., Istefanopulos, Y., Kaynak, O.: On the stability of discrete-time sliding mode control systems. IEEE Trans. Autom. Control **32**(10), 930–932 (1987)
3. Gao, W., Wang, Y., Homaifa, A.: Discrete-time variable structure control systems. IEEE Trans. Ind. Electron. **42**(2), 117–122 (1995)
4. Furuta, K.: Sliding mode control of discrete system. Syst. Control Lett. **14**(2), 145–152 (1990)

5. Drakunov, S.V., Utkin, V.I.: On discrete-time sliding modes. In: IFAC Conference on Nonlinear Control System Design. Capri, Italy (1989)
6. Utkin, V.I.: Variable Structure and Lyapunov Control. In: Sliding Mode Control in Discrete-Time and Difference Systems. Springer, Berlin (1993)
7. Hui, S., Zak, S.H.: On discrete-time variable structure sliding mode control. Syst. Control Lett. **38**(4), 283–288 (1999)
8. Brogliato, B., Acary, V., Orlov, Y.: Chattering-free digital sliding-mode control with state observer and disturbance rejection. IEEE Trans. Autom. Control **57**(5), 1087–1101 (2012)

Chapter 9
Adaptive SMC

Abstract In this chapter, the basic ideas of adaptive SMC are presented. They are intended to minimize an undesired chattering effect during motion in sliding mode. Two main approaches for designing adaptive SMC are discussed: the σ-adaptation and the dynamic adaptation based on the equivalent control, measured by a first-order low-pass filter with the discontinuous control as an input. Super-twisting control with adaptation is also considered.

Keywords Adaptive SMC · σ-adaptation · Dynamic adaptation based on the equivalent control · Adaptive super-twisting controller

The main obstacles for application of sliding mode control are two interconnected phenomena: chattering and high activity of control action. It is well known that the amplitude of chattering is proportional to the magnitude of a discontinuous control. These two problems can be handled simultaneously if the magnitude is reduced to a minimal admissible level defined by the conditions for sliding mode to exist. The adaptation methodology for obtaining the minimum possible value of control is based on two approaches developed in recent publications [1]:

- *The σ-adaptation*, providing the adequate adjustments of the magnitude of a discontinuous control within the "reaching phase", that is, when state trajectories are out of a sliding surface [2];
- *Dynamic adaptation* or the adaptation within a sliding mode (on a sliding surface), based on the *equivalent control* obtained by the direct measurements of the output signals of a first-order low-pass filter containing the discontinuous control in the input with the specially adapted magnitude value [1, 3].

9.1 The σ-Adaptation Method

Consider the nonlinear uncertain system

$$\left. \begin{array}{l} \dot{x}\,(t) = f\,(x\,(t)) + g\,(x\,(t))\,u, \\ x\,(0) = x_0, \ t \geq 0, \end{array} \right\} \tag{9.1}$$

where $x(t) \in \mathcal{X} \subset R^n$ the state vector and $u \in R$ the control input to be designed. Functions $f(x)$ and $g(x)$ are supposed to be smooth uncertain functions and are bounded for all $x \in \mathcal{X}$; furthermore, $f(x)$ contains unmeasured perturbations term and $g(x) \neq 0$ for all $x \in \mathcal{X}$.

The control objective consists in forcing the continuous function $\sigma(t, x)$, named sliding variable, to 0. Supposing that σ admits the relative degree equal to 1 with respect to u, one gets

$$\dot{\sigma}(t, x) = \Psi(t, x) + \Gamma(t, x) u,$$
$$\Psi(t, x) := \frac{\partial \sigma(t, x)}{\partial t} + \left(\frac{\partial \sigma(t, x)}{\partial x}\right)^{\mathsf{T}} f(x),$$
$$\Gamma(t, x) := \left(\frac{\partial \sigma(t, x)}{\partial x}\right)^{\mathsf{T}} g(x). \tag{9.2}$$

Functions $\Psi(t, x)$ and $\Gamma(t, x)$ are supposed to be bounded such that for all $x \in \mathcal{X}$ and all $t \geq 0$

$$|\Psi(t, x)| \leq \Psi_M, \ 0 < \Gamma_m \leq \Gamma(t, x) \leq \Gamma_M. \tag{9.3}$$

It is assumed that Ψ_M, Γ_m, and Γ_M exist but are *not known*. The objective for a designer is to propose a sliding mode controller $u(t, \sigma)$ with the same features as classical SMC, namely, robustness and finite-time convergence to some zone, closed to an equilibrium point, but without any information on uncertainties and perturbations (appearing in $f(x)$).

In the sequel, the definitions of *ideal* and *real* sliding mode are recalled.

Definition 9.1 We say that an *ideal sliding mode* exists if

$$\lim_{\sigma \to +0} \dot{\sigma} < 0 \text{ and } \lim_{\sigma \to -0} \dot{\sigma} < 0.$$

If, due to some small positive parameter μ_0, the state trajectories belong to domain

$$|\sigma(t)| \leq \Delta(\mu_0), \ \lim_{\mu_0 \to 0} \Delta(\mu_0) = 0,$$

then the motion is called a **real sliding mode**.

In real applications, an "ideal" sliding mode, as in Definition 9.1, cannot be established. But the concept of a "real" sliding mode seems to be acceptable.

The control is selected as a discontinuous state function with adaptation of its magnitude:

$$u(t, \sigma) = -K(t) \operatorname{sign}(\sigma(t, x(t))),$$
$$\dot{K}(t) = \begin{cases} \bar{K} |\sigma(t, x(t))| \operatorname{sign}(|\sigma(t, x(t))| - \varepsilon) \text{ if } K > \mu_0 \\ 0 \text{ if } K \leq \mu_0 \end{cases} \tag{9.4}$$

with $\bar{K} > 0$, $\varepsilon > 0$ and a small enough positive μ_0. The parameter μ_0 is introduced in order to get only positive values for K.

Once sliding mode with respect to $\sigma(t, x(t))$ is established, the proposed gain adaptation law (9.4) allows the gain K declining (while $|\sigma(x(t), t)| < \varepsilon$). In other words, the gain K will be kept at the smallest level that allows a given accuracy of σ-stabilization. Of course, as described in the sequel, this adaptation law allows to get an adequate gain with respect to uncertainties'/perturbations' magnitude.

Theorem 9.1 ([2]) *Given the nonlinear uncertain system (9.1) with the sliding variable $\sigma(x(t), t)$ dynamics (9.2) controlled by (9.4), there exists a finite time t_f so that a real sliding mode is established for all $t \geq t_F$, i.e.,*

$$|\sigma(x(t), t)| < \delta$$

for all $t \geq t_f$ with

$$\delta = \sqrt{\varepsilon^2 + \Psi_M^2 / \left(\bar{K}\Gamma_m\right)}. \tag{9.5}$$

So, the convergence to the domain $|\sigma| \leq \varepsilon$ is in a finite time but could be sustained in the bigger domain $|\sigma| \leq \delta$. Therefore, the real sliding mode exists in the domain $|\sigma| \leq \delta$. The choice of parameter ε has to be made by an adequate way because a "bad" tuning could provide either instability and control gain increasing to infinity or bad accuracy for closed-loop system. [2] suggested to select ε adjusted in time.

9.2 The Dynamic Adaptation Based on the Equivalent Control Method

9.2.1 The Simple Motivating Example

Consider the first-order system

$$\left.\begin{array}{l} \dot{x}(t) = a(t) + u, \\ u = -k\text{sign}(x(t)), \quad k > 0. \end{array}\right\} \tag{9.6}$$

The ranges of a time-varying parameter

$$0 < |a(t)| \leq a_+$$

and the upper bound A for its time derivative

$$|\dot{a}(t)| \leq A$$

are known only. The sliding mode with $x(t) \equiv 0$ exists for all values of unknown parameter $a(t)$ if $k > a_+$. However, if parameter $a(t)$ is varying, the gain k can be decreased and, as a result, chattering amplitude can be reduced. The objective of adaptation is decreasing k to the minimal value preserving sliding mode, if parameter a is unknown. If the condition $k > a_+$ holds, then sliding mode with $x(t) \equiv 0$ occurs and control in (9.6) should be replaced by the *equivalent control* u_{eq} for which the right-hand side in (9.6) is equal to zero:

$$\dot{x}(t) = 0 = a(t) + u_{eq}, \tag{9.7}$$

that leads to

$$k(t)\left[\text{sign}(x(t))\right]_{eq} = a(t). \tag{9.8}$$

If $k < a$, the set $x(t) \equiv 0$ is of zero measure in time and can be disregarded. The function $\left[\text{sign}(x(t))\right]_{eq}$ is an average value, or a slow component of discontinuous function $\text{sign}(x(t))$ switching at high frequency and can be easily obtained by a low-*pass filter filtering out the high-frequency component* [4]. Of course, the average value is in the range $(-1, 1)$. Then the design idea of adaptation looks natural: *after sliding mode occurs the control parameter* $\left[\text{sign}(x(t))\right]_{eq}$ *should be increased until it becomes close to* 1. On one hand, the condition $k(t) > a(t)$ should hold. But the chattering amplitude is proportional to $k(t)$. The objective of adaptation process looks now transparent:

The gain $k(t)$ should tend to $a(t)/\alpha$ **with** $\alpha \in (0, 1)$ **which is very close to 1.** As a result, the minimal possible value of discontinuity magnitude is found for the current value of parameter $a(t)$ to reduce the amplitude of chattering. For that purpose, select the *adaptation algorithm* in the form

$$\left.\begin{aligned}
\dot{k}(t) &= \rho k(t)\text{sign}(\delta(t)) - M\left[k(t) - k^+\right]_+ + M[\mu_0 - k(t)]_+ \\
\delta(t) &:= \left|\left[\text{sign}(x(t))\right]_{eq}\right| - \alpha, \ \alpha \in (0, 1) \\
[z]_+ &:= \begin{cases} 1 & \text{if } z \geq 0 \\ 0 & \text{if } z < 0 \end{cases}, \ M > \rho k^+, \ k^+ > a^+, \ \rho > \frac{A}{\alpha\mu_0} > 0.
\end{aligned}\right\} \tag{9.9}$$

The gain k can vary in the range $[\mu_0, k^+]$, $\mu_0 > 0$ is a preselected minimal value of k. For the adaptation algorithm (9.9), sliding mode will occur after a finite time interval. Indeed, if it does not exist, then

$$\left|\left[\text{sign}(x(t))\right]_{eq}\right| = 1$$

that leads to $\delta > 0$, and the increasing gain $k(t)$ will reach the value k^+ which is sufficient for enforcing sliding mode for any value of parameter $a(t)$. In sliding mode, the adaptation process (9.9) with $\delta(t) = 0$ is over after a finite time t_f. Indeed, calculate the time derivative of the Lyapunov function $V(\delta) = \delta^2/2$ assuming that during the adaptation process $k(t) \in [\mu_0, k^+]$ which means that $|a(t)|/\alpha > \mu_0$ or $(|a(t)| > \alpha\mu_0)$. The time derivatives of $\left|\left[\text{sign}(x(t))\right]_{eq}\right|$ (9.8) and $|a(t)|$ exist, and the terms depending on M in the adaptation algorithm (9.9) are equal to zero. Calculate the time derivative of the Lyapunov function $V(\delta)$ by virtue of (9.8) and (9.9):

$$
\left.
\begin{aligned}
\dot{V}(\delta) &= \delta\dot{\delta} = \delta\frac{d}{dt}\left|\left[\text{sign}(x)\right]_{eq}\right| = \delta\frac{d}{dt}\left(|a|/k\right) = \\
&\quad -|a|\,\delta k^{-1}\rho\,\text{sign}\left(\delta - M\left[k - k^+\right]_+ + M\left[\mu_0 - k\right]_+\right) \\
&\quad + \delta k^{-1}\dot{a}\,\text{sign}(a) \leq -|a|\,\delta k^{-1}\rho\,\text{sign}(\delta) + |\delta|\,k^{-1}A \\
&\quad \leq -\alpha\mu_0\rho k^{-1}|\delta| + |\delta|\,k^{-1}A = -|\delta|\,k^{-1}\left(\alpha\mu_0\rho - A\right)
\end{aligned}
\right\}
\tag{9.10}
$$

and if $\rho > A/\alpha\mu_0$ it follows

$$
\dot{V}(\delta) \leq -\sqrt{2}\frac{(\alpha\mu_0\rho - A)}{k^+}\sqrt{V(\delta)},
$$

implying that $\sqrt{V(\delta)} = 0$ at least after

$$
t_f = \frac{k^+}{(\alpha\mu_0\rho - A)}\sqrt{2V(\delta(0))} = \frac{k^+}{(\alpha\mu_0\rho - A)}|\delta(0)|
$$

and, as a result, $\delta(t)$ becomes equal to zero identically after the finite time t_f.

After the adaptation process is over $(t > t_f)$, we have

$$
\left|\left[\text{sign}(x(t))\right]_{eq}\right| = \frac{|a|}{k} - \alpha.
$$

So, $k = |a|/\alpha$. If in the course of motion $|a(t)|/\alpha < \mu_0$, then the gain $k(t)$ decreases until $k(t) = \mu_0$ and, as it follows from (9.9), it will be maintained at this level. Since the gain $a(t)$ is time varying, its increase can result in $|a(t)|/\alpha = \mu_0$ and $\delta(t) = 0$ at a time t_f. As it follows from the above analysis, for the further motion in the domain $k(t) \in (\mu_0, k^+]$ with the initial condition $\delta(t_f) = 0$ the time function $\delta(t)$ will be equal to zero with $\alpha = |a(t)|/k(t)$.

Remark 9.1 The function $\left[\text{sign}(x(t))\right]_{eq}$ is needed here for the implementation of the adaptation algorithm (9.9). It can be derived by filtering out a high-frequency component of the discontinuous function $\text{sign}(x(t))$ by a low-pass filter

$$
\tau\dot{z} + z = \text{sign}(x(t)), \quad z(0) = 0
$$

with a small time constant $\tau > 0$ and the output $z(t, \tau)$ which is, in fact, an estimate of $\left[\text{sign}(x(t))\right]_{eq}$ satisfying

$$
\left|z(t, \tau) - \left[\text{sign}(x(t))\right]_{eq}\right| \leq O(\tau) \underset{\tau\to 0}{\longrightarrow} 0.
$$

Then the convergence analysis of (9.9) and (9.10) with $\delta(t, \tau) = z(t, \tau) - \alpha$ is valid beyond the domain $|\delta(t, \tau)| \leq O(\tau)$. This inequality defines the accuracy of adaptation. Note that the switching frequencies of the modern power converters are of order dozens of kHz, and very small time constant τ can be selected to get a high accuracy of adaptation.

Notice also that, as follows from [4],

$$z(t, \tau) = \left[\text{sign}(x(t)) \right]_{eq} + \psi(t) + O(\sup|x(t)| + \tau) + O(\tau^{-1}\sup|x(t)|)$$

where $\psi(t)$ is the fast rate exponentially decreasing function. The term $\sup|x(t)|$ is inverse proportional to the sliding mode frequency f. It is of order of dozen kHz in the modern switching devices. Therefore, it is not a problem to make the term

$$O(\sup|x| + \tau) + O(\tau^{-1}\sup|x|)$$

negligible. Of course, this engineering language can be translated into mathematical one, for example, as follows: *for any $\varepsilon > 0$,   there exists a switching frequency f_0 such that*

$$|z(t, \tau) - u_{eq}| < \varepsilon,$$

if $f > f_0$ implying

$$\text{sign}[\left| \left[\text{sign}(x(t)) \right]_{eq} \right| - \alpha] = \text{sign}[|z(t, \tau)| - \alpha].$$

9.2.2 Multidimensional Case

9.2.2.1 Main Assumptions

Here, we consider an arbitrary order system

$$\left. \begin{array}{c} \dot{x}(t) = f(t, x(t)) + b(t, x(t)) u(t, x(t)) \\ x(t) \in R^n, \ f : R^+ \times R^n \to R^n \\ u : R^+ \times R^n \to R, \ b : R^+ \times R^n \to R^n \end{array} \right\} \quad (9.11)$$

for which we assume that

A1 the control $u = u(t, x)$ enforces siding mode on some surface $\sigma(x) = 0$ ($\sigma \in C^1$) and is in the following form:

$$\left. \begin{array}{c} u(t, x) = -k(t) \left(1 + \lambda\sqrt{\|x\|^2 + \varepsilon} \right) \text{sign}(\sigma(x)) \\ \lambda \geq 0, \ \varepsilon > 0, \ k(t) \in \left[\mu_0, k^+ \right], \ \mu_0 > 0 \end{array} \right\} \quad (9.12)$$

Similar to example (9.6), the control gain $k(t)$ is a time-varying function governed by the adaptation procedure described below.

A2 the uncertain functions $f(t, x)$ and $b(t, x)$ satisfy the commonly accepted conditions (which are much more general than in (9.3)):

$$\left. \begin{array}{c} \|f(t, x)\| \leq f_0 + f_1 \|x\|, \ 0 < b_0 \leq \nabla^\mathsf{T} \sigma(x) b(t, x), \\ \|b(t, x)\| \leq b^+, \ \|\nabla \sigma(x)\| \leq \sigma^+ \end{array} \right\} \tag{9.13}$$

$$\Phi(t, x) := \frac{\nabla^\mathsf{T} \sigma(x) f(t, x)}{\nabla^\mathsf{T} \sigma(x) b(t, x)},$$

$$\|\nabla^\mathsf{T} \Phi(t, x)\| \leq \Phi_0 + \Phi_1 \|x\|, \ \left| \frac{\partial}{\partial t} \Phi(t, x) \right| \leq \varphi_0 + \varphi_1 \|x\|. \tag{9.14}$$

All coefficients in the right-hand sides of these inequalities are constant and positive. The function $\sigma(x)$ and its time derivative

$$\left. \begin{array}{c} \dot\sigma(x) = \nabla^\mathsf{T} \sigma(x) f(t, x) - \\ \nabla^\mathsf{T} \sigma(x) b(t, x) k(t) \left(1 + \lambda\sqrt{\|x\|^2 + \varepsilon} \right) \text{sign}(\sigma(x)) \end{array} \right\} \tag{9.15}$$

should have opposite signs ($\sigma(x)\dot\sigma(x) < 0$ if $\sigma(x) \neq 0$), which is the condition for sliding mode to exist on the surface $\sigma(x) = 0$. The sufficient condition for this follows from (9.13), (9.14), and (9.15):

$$\sigma(x)\dot\sigma(x) = \sigma(x)\nabla^\mathsf{T}\sigma(x) f(t, x) -$$
$$\nabla^\mathsf{T}\sigma(x) b(t, x) k(t) \left(1 + \lambda\sqrt{\|x\|^2 + \varepsilon} \right) |\sigma(x)| \leq$$
$$[\nabla^\mathsf{T}\sigma(x) b(t, x)] |\sigma(x)| \left[|\Phi(t, x)| - k(t) \left(1 + \lambda\sqrt{\|x\|^2 + \varepsilon} \right) \right] < 0,$$

if

$$|\Phi(t, x)| - k(t) \left(1 + \lambda\sqrt{\|x\|^2 + \varepsilon} \right) < 0 \tag{9.16}$$

which always holds when

$$\lambda \geq f_1/f_0, \ \mu_0 > f_0 \sigma^+/b_0, \ k(t) \in \left(\mu, k^+ \right] \tag{9.17}$$

in view of the relation

$$|\Phi(t, x)| - k(t) \left(1 + \lambda\sqrt{\|x\|^2 + \varepsilon} \right) \leq f_0 \frac{\sigma^+ (1 + \|x\| f_1/f_0)}{b_0} - \mu_0 (1 + \lambda\|x\|).$$

To derive the sliding mode equation, the function sign($\sigma(x)$) should be replaced by the solution of the equation $\dot\sigma(x) = 0$ with respect to the term sign($\sigma(x)$), called *the equivalent control*:

$$\left[\text{sign}\,(\sigma\,(x)) \right]_{eq} :=$$

$$\begin{cases} \dfrac{\varPhi\,(t, x)}{k\,(t)\left(1 + \lambda\sqrt{\|x\|^2 + \varepsilon}\right)} & \text{if} \quad \sigma\,(x\,(t)) = 0 \\[2mm] \text{sign}\,(\sigma\,(x\,(t))) & \text{if} \quad \sigma\,(x\,(t)) \neq 0 \end{cases} \tag{9.18}$$

satisfying (in view of (9.16)) in the sliding mode ($\sigma\,(x\,(t)) = 0$)

$$\left| \left[\text{sign}\,(\sigma\,(x)) \right]_{eq} \right| < 1. \tag{9.19}$$

9.2.2.2 Description of the Adaptation Procedure

The idea of the *adaptation law* for the control gain $k\,(t)$ is similar to that for our first-order system in the previous subsection:

$$\dot{k}(t) = \begin{cases} (\gamma_0 + \gamma_1\,\|x\|)\,k(t)\text{sign}\,(\delta\,(t)) \\ -\,M\left[k(t) - k^+\right]_+ + M\,[\mu_0 - k(t)]_+ \end{cases} \tag{9.20}$$

where

$$\delta(t) := \left| \left[\text{sign}\,(\sigma\,(x)) \right]_{eq} \right| - \alpha \tag{9.21}$$
$$\alpha \in (0, 1)\,, \ \lambda > 0,\ \gamma_0, \gamma_1 > 0.$$

Select in (9.20) $k^+ > \sigma^+ \dfrac{f_0}{b_0}$. If sliding mode does not exist, then

$$\left| \left[\text{sign}\,(\sigma\,(x)) \right]_{eq} \right| = 1$$

and the gain $k(t)$ will be equal to k^+ which results in the occurrence of this motion in the surface $\sigma\,(x\,(t)) = 0$.

Theorem 9.2 ([3]) *For the dynamic system (9.11) closed by the control (9.12) with the gain adaptation law (9.20) and (9.21) with the parameters satisfying*

$$\left.\begin{array}{c} k^+ > \sigma^+ \dfrac{f_0}{b_0},\ \mu_0 > f_0\sigma^+/b_0,\ 0 < \varepsilon \ll 1 \\[3mm] \gamma_0 > \alpha^{-1}\left[\left(\dfrac{f_0}{\mu_0} + b^+\right)\varPhi_0 + \dfrac{\varphi_0}{\mu_0} + f_0 + b^+ k^+\right] \\[3mm] \gamma_1 \geq \alpha^{-1}\left(\dfrac{f_0}{\mu_0} + b^+\right)\varPhi_1,\ M > \gamma_0 k^+ \end{array}\right\} \tag{9.22}$$

there exist

$$\theta := \alpha\gamma_0 - \left[\left(\frac{f_0}{\mu_0} + b^+\right)\Phi_0 + \frac{\varphi_0}{\mu_0} + f_0 + b^+ k^+\right] > 0 \tag{9.23}$$

and $t_f = \theta^{-1}|\delta(0)|$ (where $\delta(0)$ is defined by (9.21)) such that for all $t \geq t_f$ the condition

$$\left|\left[\text{sign}\,(\sigma\,(x\,(t)))\right]_{eq}\right| = \alpha \tag{9.24}$$

holds. It means that the sliding surface $\sigma(x) = 0$ is attained in a finite time t_f, and for small enough positive number ε_0 the suggested adaptation procedure with $\alpha = 1 - \varepsilon_0$ provides $k(t)$ tending to a vicinity of the minimum possible value $k_{min}(t)$, that is, as it follows from (9.18), in sliding mode

$$k(t) = \begin{cases} \dfrac{1}{1-\varepsilon_0}k_{min}(t) & if\ k_{min}(t) \geq \mu \\ \mu & if\ k_{min}(t) < \mu \end{cases} \tag{9.25}$$

$$k_{min}(t) := \frac{|\Phi(t, x(t))|}{1 + \lambda\sqrt{\|x(t)\|^2 + \varepsilon}}.$$

9.3 Super-Twisting Control with Adaptation

The dynamic system with super-twist controller can be represented as

$$\begin{aligned} \dot{x}_1 &= x_2 - \bar{\alpha}\sqrt{|x_1|}\text{sign}(x_1), \\ \dot{x}_2 &= \phi(t) - k\text{sign}(x_1). \end{aligned} \tag{9.26}$$

Take $\sigma(x) = x_1$ and permit for the gain parameter to be time-varying, i.e., $k(t) = \beta(t)$.

Theorem 9.3 (on adaptive super-twisting [3])
The system (9.26) with disturbances $\phi(t)$ having a bounded derivative (fulfilling $\frac{d}{dt}|\phi(t)| \leq L$), and with the parameter $\bar{\beta}(t) = k(t)$ adapted online according to the adaptation law

$$\dot{k}(t) = \begin{cases} \gamma_0 k(t)\text{sign}\,(\delta(t)) - M\left[k(t) - k^+\right]_+ + M\left[\mu_0 - k(t)\right]_+ \\ \qquad if\ 0 < \mu_0 \leq k(t) \leq k^+ \\ \\ 0\ otherwise \end{cases}$$

where $\gamma_0 > L/\mu$ converges in the finite time

$$t_f = |\delta(0)|\,k^+/(\mu_0\gamma_0 - L)$$

to the sliding mode regime $\sigma(x) = x_1 = 0$ maintaining within the relation

$$\phi(t)/k(t) = \alpha = 1 - \varepsilon_0$$

for small enough $\varepsilon_0 > 0$. It means that the magnitude of the discontinuous control is close to its minimal value as well as the amplitude of chattering.

References

1. Utkin, V.I., Poznyak, A.S.: Advances in Sliding Mode Control: Concept, Theory and Implementation. In: Adaptive Sliding Mode Control. Springer, Berlin (2013a)
2. Plestan, F., Shtessel, Y., Bregeault, V., Poznyak, A.: New methodologies for adaptive sliding mode control. Int. J. Control **83**(9), 1907–1919 (2010)
3. Utkin, V.I., Poznyak, A.S.: Adaptive sliding mode control with application to super-twist algorithm: Equivalent control method. Automatica **49**(1), 39–47 (2013a)
4. Utkin, V.: Sliding Modes in Control and Optimization. Springer, Berlin (1992)

Chapter 10
SMC in Infinite-Dimensional Systems

Abstract This brief chapter introduces a reader in the problem of SMC designing the class of infinite-dimensional systems governed by a linear parabolic PDE containing uncertainties. The distributed unit signal is considered as a robust controller. The Lyapunov-Krasovskii functional is suggested for the analysis of the corresponding closed-loop dynamics.

Keywords Infinite-dimensional systems · Parabolic PDE · Distributed unit signal · Lyapunov-Krasovskii functional

Many important applications are governed by functional and partial differential equations or, more general, differential equations in a Hilbert space. Relevant examples are flexible manipulators and structures, heat transfer processes, combustion, and fluid mechanical systems, as well as retarded systems. As these systems are often described by models with a significant degree of uncertainty, an interest has emerged to develop consistent control methods that would be capable of utilizing distributed-parameter and time-delay models and providing the desired system performance in spite of significant model uncertainties.

Thus motivated, SMC algorithms have extensively been generalized in the infinite-dimensional setting starting from early 1980s. First infinite-dimensional generalizations of SMC algorithms, made in [1–3] for some parabolic and hyperbolic PDEs, not only corroborated their utility but also faced with challenging problems calling for further investigation.

10.1 Infinite-Dimensional Setting

SM challenges, arising in the infinite-dimensional setting, can readily be illustrated by a benchmark example of the heat diffusion along a 1-D bar with isolated ends. Such a diffusion process is governed by the linear parabolic PDE

$$x_t(r, t) = x_{rr}(r, t) + f(r, t) + u(r, t) \qquad (10.1)$$

with the following Neumann boundary conditions:

$$x_r(0, t) = 0, \quad x_r(1, t) = 0, \tag{10.2}$$

where $x(\cdot, t)$ is the temperature distribution along the bar at a time instant $t > 0$, and the control input u and external disturbance f affect the bar temperature through the entire spatial bar location $r \in (0, 1)$. Hereinafter, the symbols

$$x_t(r, t) = \frac{\partial x(r, t)}{\partial t}, \quad x_r(r, t) = \frac{\partial x(r, t)}{\partial r}, \quad x_{rr}(r, t) = \frac{\partial^2 x(r, t)}{\partial r^2}$$

stand for the corresponding partial state derivatives, and the symbol $\|x(\cdot, t)\|_2$ is reserved for the L_2-norm $\sqrt{\int_0^1 x^2(r, t)\, dr}$.

Let the distributed unit signal

$$u(r, t) = -M \frac{x(r, t)}{\|x(\cdot, t)\|_2}, \tag{10.3}$$

magnified by the gain $M > 0$, is applied to the heat process (10.1) for enforcing the plant dynamics to slide the plant dynamics along the discontinuity set $x = 0$. By inspection, the L_2-norm of the unit signal $x(r, t)\|x(\cdot, t)\|_2^{-1}$ does equal one everywhere but the origin $x \equiv 0$. Fundamental questions then arise:

- Under which conditions a sliding mode occurs on the discontinuity set?
- What equivalent value of the unit signal $x(r, t)\|x(\cdot, t)\|_2^{-1}$ should be utilized in the boundary-value problem (10.1), (10.2) to correctly describe the sliding mode on the discontinuity set $x = 0$?

Being inspired by the finite-dimensional counterpart, the conditions

$$\lim_{x \uparrow 0} x_t(r, t) = \lim_{x \uparrow 0} [x_{rr}(r, t) + f(r, t) + u(r, t)] > 0,$$
$$\lim_{x \downarrow 0} x_t(r, t) = \lim_{x \downarrow 0} [x_{rr}(r, t) + f(r, t) + u(r, t)] < 0 \tag{10.4}$$

for sliding dynamics to locally exist along the discontinuity set $x(\cdot, t) = 0$ require the controller magnitude M to exceed (locally in t) not only the uniform norm $\|f(\cdot, t)\|_{C(0,1)}$ of the external disturbance, but also the uniform norm $\|x_{rr}(\cdot, t)\|_{C(0,1)}$ of the second-order spatial derivative x_{rr}. The latter requirement is, however, impossible to meet because the spatial double differentiation operator $A = \dfrac{\partial^2}{\partial r^2}$ is unbounded (indeed, $\|A \cos(nr)\|_{C(0,1)} \to \infty$ as $n \to \infty$). Thus, the SM existence condition (10.4) should be revised in the infinite-dimensional setting.

The Lyapunov-Krasovskii (Lyapunov-Razumikhin) approach is presently well recognized for revealing SMs in the PDEs (respectively, delayed functional differential equations) setting [4]. In a particular case of the closed-loop system (10.1)–(10.3), a Lyapunov-Krasovskii functional can be chosen in the form

$$V(x) = \frac{1}{2} \int_0^1 x^2(r, t) dr. \tag{10.5}$$

Indeed, differentiating (10.5) along the solutions of (10.1)–(10.3) yields

$$\dot{V} = \int_0^1 [x_{rr} + u + f] x \, dr \leq -(M - \|f(\cdot, t)\|_{L_2}) \|x(\cdot, t)\|_{L_2}$$

$$= -(M - \|f(\cdot, t)\|_{L_2}) \sqrt{2V(t)}. \tag{10.6}$$

Thus, provided that

$$M > \|f(\cdot, t)\|_{L_2},$$

the finite-time stability of the closed-loop system (10.1)–(10.3) is established and the existence of sliding modes on the discontinuity set $x = 0$ is guaranteed in finite time

$$T = \sqrt{2}(M - \|f(\cdot, t)\|_{L_2})^{-1} \sqrt{V(0)}.$$

10.2 Unit Control for Infinite-Dimensional Systems

Generalization of the unit control design for arbitrary systems in Hilbert space is further presented. For infinite-dimensional dynamics

$$\dot{x} = Ax + Bu + f(t, x),$$

evolving in a Hilbert space and affected by unknown disturbances, the typical control objective is their asymptotical stabilization. The state and control are elements of Hilbert spaces $x \in X, u \in U, f : R \times X \to X$ is a disturbance,

$$A : D(A) \subset X \to X$$

is a closed linear operator with the domain $D(A)$ dense in X and $B : U \to X$ is a linear bounded operator, the unknown disturbance $f(t, x)$ is assumed to be bounded with an *a priori* known norm bound. Of course, the problem can be solved if the matching condition

$$f(t, x) = B\gamma(t, x), \|\gamma(t, x)\| \leq \gamma_0, \gamma : R \times X \to U$$

holds.

Without touching details on strong solutions of the closed-loop system to exist (the interested reader can refer, e.g., to [4, 5]), select the design procedure, following the methodology, offered in Sect. 3.8 and in [6, 7]. Such a procedure becomes applicable if the nominal system $\dot{x} = Ax$ is exponentially stable with a known positive-definite continuous Lyapunov functional

$$v : X \to R, \, v(0) = 0, \, v(x) > 0 \text{ for } x \neq 0$$

such that

$$\dot{v} = \nabla_x\{v\}Ax < 0,$$

where $\nabla_x\{v\} : X \to R$ is a linear functional similar to $\nabla^\top V : R^n \to R$ for finite-dimensional space with the property

$$\lim_{\|\Delta x\| \to 0} \frac{v(x + \Delta x) - v(x) - \nabla_x\{v\}(\Delta x)}{\|\Delta x\|} = 0, \quad \Delta x \in X.$$

Calculating the time derivative of $v(x)$ on trajectories of the original systems yields

$$\dot{v} = \nabla_x\{v\}Ax + \nabla_x\{v\}B(u + \gamma).$$

Let

$$(\nabla_x\{v\}B)^* \in U$$

be an adjoint operator to $\nabla_x\{v\}B$ [8]. By property of an adjoint operator [9],

$$\nabla_x\{v\}B \, (\nabla_x\{v\}B)^* = \|\nabla_x\{v\}B\|^2.$$

For control in the form

$$u = -\rho \frac{(\nabla_x\{v\}B)^*}{\|\nabla_x\{v\}B\|}, \quad \rho > \gamma_0$$

(with the same motivation as before in Sect. 3.8 it can be called unit control), we derive

$$\dot{v} \leq \nabla_x\{v\}Ax + \|\nabla_x\{v\}B\|(-\rho + \gamma_0) < 0.$$

Hence, the state $x = 0$ is asymptotically stable provided that strong solutions of the closed-loop system exist.

Unit control approach in the infinite-dimensional setting has successfully been illustrated for the parabolic boundary-value problem (10.1), (10.2). Note that along with the unit control input (10.3), the distributed relay control $u = -\rho\text{sign}(x)$ of the diffusion process results in the asymptotic stabilization as well, but an upper estimate of the decay of $x(t, r)$ has not been found until now. Even a solution concept for the closed-loop diffusion system, driven by such a distributed relay input, should be revisited. In addition, the unit control is a continuous function during the transient process and becomes discontinuous only when the process is over and x becomes equal to zero for any point r. It is not the case for the distributed relay control, undergoing discontinuity at the points, where $x(t, r) = 0$.

It is worth noticing that in the above example, the sliding mode set $x = 0$ manifests the trivial sliding mode dynamics $\dot{x} = 0$ itself. However, in order to describe sliding mode dynamics in less trivial examples where sliding modes appear to be not confined to the origin, the equivalent control method should properly be revalidated in the infinite-dimensional setting. Infinite-dimensional extensions of the equivalent control method, capturing a broad class of semilinear PDEs, were developed in [10, 11]. These extensions, coupled to the Lyapunov-Krasovskii and Lyapunov-Razumikhin approaches, resulted in numerous SMC algorithms of the infinite-dimensional systems. Relevant references may be found in [4] for Distributed-Parameter Systems (DPS) and in [12] for time-delay systems.

Recently, a powerful backstepping tool was extended to the PDEs setting [13] through a non-trivial integral state transformation for the purpose of stabilizing linear DPS of infinite relative degree with anticollocated boundary sensing and actuation. This technique was then combined with SMC algorithms in order to robustify the existing PDEs-oriented boundary controllers. Such combined SM-backstepping algorithms were proposed in [14–16] to name a few.

The modern trend in the DPS design is twofold. The closed-loop DPS is required to possess a good performance not only in the L_2 state space, but also in a Sobolev space. Having a good performance in the former would admit destroying peaks within small spatial subdomains of the operation. Apart from this, having a good performance in a Sobolev space would result in the uniform point-wise state decay, thereby avoiding extremely undesired peaking phenomena within the spatial operation domain. Motivated by this trend, twisting and super-twisting algorithms were recently re-worked for uncertain heat and wave propagation processes (see [17] and references therein) to address their point-wise stabilization and observation where the available sensing and actuation were either distributed over the entire state space or located at the boundary. Certainly, the SM theory of the infinite-dimensional systems is in progress, and many SMC challenges remain open in the infinite-dimensional setting and promising directions in their further investigation are announced in the next section.

References

1. Orlov, Y., Utkin, V.I.: Use of sliding modes in distributed system control problems. Autom. Remote Control **43**, 1127–1135 (1982)
2. Orlov, Y.: Application of Lyapunov method in distributed systems. Autom. Remote Control **44**, 426–430 (1983)
3. Zolezzi, T.: Differential inclusions and sliding mode control. In: Sliding Mode Control in Engineering (W. Perruquetti and J.P. Barbot, Eds.). Marcel Dekker, pp. 29–52 (2002)
4. Orlov, Y.: Discontinuous Systems: Lyapunov Analysis and Robust Synthesis Under Uncertainty Conditions. Springer, Berlin (2008)
5. Curtain, R., Zwart, H.: An Introduction to Infinite-dimensional Linear Systems. Springer, New York (1995)
6. Gutman, S.: Uncertain dynamic systems—a lyapunov min-max approach. IEEE Trans. Autom. Control **AC-24**, 437–449 (1979)

7. Gutman, S., Leitmann, G.: Stabilizing feedback control for dynamic systems with bounded uncertainties. In: Confrence on Decision and Control, pp. 94–99 (1976)
8. Orlov, Y., Utkin, V.: Unit sliding mode control in infinite-dimensional systems. J. Appl. Math. Comput. Sci. **8**, 7–20 (1998)
9. Rudin, W.: Functional Analysis, 2nd edn. McGraw-Hill, New York (1991)
10. Orlov, Y., Dochain, D.: Discontinuous feedback stabilization of minimum-phase semilinear infinite-dimensional systems with application to chemical tubular reactor. IEEE Trans. Autom. Control **47**, 1293–1304 (2002)
11. Levaggi, L.: Infinite dimensional systems sliding motions. Eur. J. Control **8**, 508–518 (2002)
12. Yan, X., Spurgeon, S., Edwards, C.: Variable-Structure Control of Complex Systems. Springer, Berlin (2017)
13. Krstic, M., Smyshlyaev, A.: Boundary Control of PDEs: A Course on Backstepping Designs. SIAM, Philadelphia PA (2008)
14. Guo, B.-Z., Jin, F.-F.: Sliding mode and active disturbance rejection control to stabilization of one-dimensional anti-stable wave equations subject to disturbance in boundary input. IEEE Trans. Autom. Control **58**(13), 1269–1274 (2013)
15. Cheng, M.B., Radisavljevic, V., Su, W.C.: Sliding mode boundary control of a parabolic pde system with parameter variations and boundary uncertainties. Automatica **47**(2), 381–387 (2011)
16. Guo, B.Z., Jin, F.-F.: Output feedback stabilization for one-dimensional wave equation subject to boundary disturbance. IEEE Trans. Autom. Control **60**, 824–830 (2015)
17. Pisano, A., Orlov, Y.: Boundary second-order sliding-mode control of an uncertain heat process with unbounded matched perturbation. Automatica **48**, 1768–1775 (2012)

Chapter 11
Open Problems in SMC

Abstract This chapter describes several open problems that can be formulated in the frame of SMC concept. They are

- stability problem of the system $\dot{x} = A\,\mathrm{sign}(x)$,
- SMC in the presence of noise in measurements,
- finite-time stability of SMs in infinite-dimensional systems,
- SMC for systems governed by shift operators, and
- stochastic sliding mode.

Keywords Stability problem · Noise in measurements · Finite-time stability of SMs in infinite-dimensional systems · Shift operators · Stochastic SM

There are many open challenging problems in each of the above chapters. Certainly, it does not exclude totally new problems which are formulated as well.

11.1 Stability Problem of the System $\dot{x} = A\,\mathbf{sign}(x)$

We hope that the stability conditions for the system

$$\dot{x}(t) = A\,\mathrm{sign}(x(t)), \tag{11.1}$$

where

$$x(t) = (x_1(t), \ldots, x_n(t))^T \in R^n,$$

and $A \in R^{n \times n}$ being a constant matrix, can be used for finding sliding domains in the system (2.7). Note that necessary and sufficient stability conditions for (11.1) were established for two-dimensional systems in [1]. The existence of sliding mode on the surface $s = 0$ (see, (3.3)) is determined by stability of the equation

$$\dot{s}(t) = -MD(t)\mathrm{sign}(s) + d(t),$$

© The Author(s), under exclusive license to Springer Nature Switzerland AG 2020
V. Utkin et al., *Road Map for Sliding Mode Control Design*,
SpringerBriefs in Mathematics,
https://doi.org/10.1007/978-3-030-41709-3_11

where $M > 0$ is a tuning parameter, $d(t)$ is a disturbance, and $D(t)$ is a time-varying matrix. We hope that stability conditions of the time-varying system can be formulated in terms of stability of time-invariant system (11.1). Then, an optimal (in some sense) selection of a sliding surface may improve significantly the control performance.

To the moment, we have only one argument in favor of our hope. Let us assume that we can find Lyapunov function as a quadratic form $V = x^T P x$ for any constant matrix A with matrix $P > 0$, depending on A, such that

$$\dot{V} = x^T P A \text{sign}(x) + \text{sign}^T(x) A^T P x < 0.$$

Let now $A(t)$ be a time-varying matrix with bounded time derivative, and let the system be "*skillfully*" stable, that is, the real part of the spectrum of matrix A is equal or less than $-\delta$, $\delta > 0$ for any fixed time t. We can find Lyapunov function

$$V(t) = x^T P(t) x > 0$$

with $P(t) > 0$ depending on time. It can be shown that $\| P(t) \|$ as well as $\| \dot{P}(t) \|$ are bounded. Calculate the derivative of $V(t)$:

$$\dot{V}(t) = x^T P(t) A(t) \text{sign}(x) + \text{sign}^T(x) A^T(t) P(t) x + x^T \dot{P}(t) x.$$

The first two terms tend to zero with x tending to zero as a linear function, whereas the third term tends to zero as a quadratic function. It means that the origin in space x is asymptotically stable for the time- varying system as well (of course it is not always the case for linear systems $\dot{x} = A(t)x$). Then, the sliding domain can be found from equation

$$\dot{s} = G(x, t) f(x, t) + G(x, t) B(x, t) u,$$

if the matrix $G(x, t) B(x, t)$ satisfies stability condition for any constant x and t.

For linear systems $\dot{x}(t) = A x(t)$, the necessary and sufficient condition of asymptotic stability of the origin is well known: all eigenvalues of the matrix A possess negative real parts. However, this is not the case for the relay system (11.1). For instance, the matrix

$$A = \begin{pmatrix} -3 & 1 & 1 \\ 4.5 & -1 & 2 \\ 4.5 & -1.5 & -1 \end{pmatrix}$$

is Hurwitz, but the corresponding relay system (11.1) is unstable. In this particular case, instability can be easily proven by means of the equivalent control method. Indeed, $x_1 = 0$ is the globally attractive sliding surface and one has

$$[\text{sign}(x_1)]_{eq} = 1/3(\text{sign}(x_2) + \text{sign}(x_3))$$

as soon as sliding mode occurs. It becomes evident that the system is unstable after substitution of the equivalent control into the third equation, since

$$\dot{x}_3 = 0.5\text{sign}(x_3).$$

Several sufficient stability conditions for (11.1) have been presented in [2–4]. The simplest one can be formulated in the form of linear matrix inequality:

$$A^T \Lambda + \Lambda A < 0,$$

where Λ is a diagonal positive-definite matrix. The necessary and sufficient condition of asymptotic stability of the relay system (11.1) is still not discovered. We consider this as the open problem, which is important for tuning of relay sliding mode control algorithms.

11.2 SMC in the Presence of Noise in Measurements

The output-based sliding mode design is usually studied without any assumption about noise in the measurement. However, in practice, the measurements are usually corrupted by noise, and an appropriate modification of the sliding mode control methodology is required.

Consider the system

$$\dot{x}(t) = Ax(t) + Bu(t) + Dg(t), \quad t > 0,$$
$$y(t) = Cx(t) + w(t), \quad x(0) \in x_0,$$

where $x(t) \in R^n$ is the system state, $u(t)$ is the control input, $y(t) \in R^k$ is the measured output, the functions $w : R \to R^k$, and $g : R \to R^p$ describe measurement noise and exogenous disturbance, respectively, the matrices

$$A \in R^{n \times n}, \ B \in R^{n \times m}, \ C \in R^{k \times n}, \ D \in R^{n \times p}$$

are assumed to be known. Note that for $Dg(t) \in \text{range}(B)$ the exogenous disturbance becomes matched.

As usual, we deal with the classical sliding mode control problem: to design a control algorithm, which realizes finite-time reaching of a given linear manifold

$$Fx = 0, \ F \in R^{m \times n}, \det(FB) \neq 0$$

and further sliding on this manifold.

Due to measurement noises and system disturbances, the sliding mode cannot appear on the given manifold $Fx = 0$. For practice, it is important to know that which sort of feedback control is optimal to provide the system motion as close as possible to the preselected manifold.

The case of L_2 bounded noises and disturbances has been studied using the minimax observer design in [5]. It was proven that a possible optimal control law realizing required motion ($\|Fx(t)\| \to$ min) is a linear feedback. Notice that in the presence of the observation noise $\omega(t)$, this minimum never can be done equal to zero (independently whether the noise is from L_2 or from L_∞). In the case of a measurement noise, the convergence of both the state estimates and the distance to the desired manifold to some zone can be guaranteed only. The following question arises: *under which restrictions to measurement noises and disturbances the SM controllers and observers provide the convergence of the system trajectories to a smaller zone than linear algorithms?* It can be considered as an interesting open problem.

Another important problem is the optimal selection of the sliding manifold $Fx = 0$ to minimize the effects of mismatched perturbations. For the noise-free state feedback sliding mode control with $y = x$, the sliding mode control problem has been studied in [6] for exogenous disturbances bounded in L_∞. The background for possible future developments in this context is presented in [7], where the general framework for L_2-gain analysis of sliding mode controllers is introduced.

11.3   Finite-Time Stability of SMs in Infinite-Dimensional Systems

The extension of the SM analysis tools to the infinite-dimensional setting is far from being trivial. Although the finite-time extinction in the infinite-dimensional setting has been recognized in the literature for (linear, in particular) evolution equations (see [8] and references therein), however, the finite-time stability analysis of SMs remains a challenging problem for infinite-dimensional systems. The strict Lyapunov functional (10.5), resulting in the finite-time stability of the closed-loop heat process (10.1), (10.2), enforced by the unit distributed control signal (10.3), remains the only one which is available in the PDE setting [9].

It should be pointed out that the conventional homogeneity approach proves to be useless in the finite-time stability analysis of the above closed-loop system (10.1), (10.2), (10.3). Indeed, the right-hand side of the PDE (10.1) contains linear and unit terms, which are of distinct homogeneity degree on one hand, so that it cannot be homogeneous in the standard sense. On the other hand, the unit term, magnified with a finite gain, cannot locally exceed the linear unbounded operator of the double differentiation so that it cannot reject (even locally) such a disturbing term. Such a situation is opposed to that in the finite-dimensional setting.

Remarkably, the extension of the homogeneity concept to infinite-dimensional systems, proposed in [10], is neither suitable for the finite-time stability analysis of the rather trivial distributed relay control example, given above, and it calls for an appropriate modification in the infinite-dimensional setting. In addition, the existing counterparts of the twisting and super-twisting algorithms, developed in the heat and wave PDEs settings [11], manifest the need for the deeper finite-time stability insight on these algorithms in the PDE setting as they have been established to yield the closed-loop asymptotic stability only.

11.4 A Little Fantasy: SMC for Systems Governed by Shift Operators

Let us consider the evolution equation [12] of the form

$$\dot{x}(t) = f(t, x(t)), \quad t > t_0, \quad x(t_0) = x_0, \tag{11.2}$$

where $x(t)$ is an element of an appropriate finite or infinite-dimensional linear space $\mathbf{X}$ and $f : R \times \mathbf{X} \to \mathbf{X}$ is an appropriate operator (function). We assume that the Cauchy problem (11.2) has a solution $p(t, t_0, x_0)$ for any $x_0 \in \mathbf{X}$.

A model of dynamic system usually considers its state $x(t)$ (or $x(k)$) as an element (point) of an appropriate finite- or infinite-dimensional normed linear space X, where t denotes a continuous time (and k denotes a discrete time). The evolution of the dynamic system in time defines a trajectory (curve) in the space X. A point $x(t_0) = x_0$ starting from an instant of time t_0 moves along the trajectory and reaches a point $x(t_1) = x_1$ at a time instant t_1. The operator

$$U(t_1, t_0) : X \to X$$

of translation of the point $x(t_0) = x_0$ to the point $x(t_1) = x_1$ is called shift operator [13] (or "solution operator"). The operator $U(t_1, t_0)$ is single-valued if the trajectory is unique. The classical example of the shift operator is the matrix exponent

$$U(t_1, t_0) = e^{A(t_1 - t_0)}$$

for a linear ODE:

$$\dot{x}(t) = Ax(t), x \in \mathbb{R}^n, A \in \mathbb{R}^{n \times n}.$$

We confine ourself to an autonomous case only. In this case, a dynamic system has an evolution law, which is invariant with respect to time, so the shift operator depends only on $t_1 - t_0$ (i.e., $U(t_1, t_0) = U(t_1 - t_0)$), so $U(s)$ is a semigroup on X, indeed, $U(0) = I$ is an identity operator and

$$U(t + s) = U(t)U(s), \quad t, s > 0$$

is a semigroup property. The shift operator may be unbounded if trajectories blow up (i.e., $\|x(t)\|_X \to +\infty$ as $t \to T_* < +\infty$). Note that if the trajectories are not unique in the backward time, then the shift operator is not invertible or, at least, its inverse is not single-valued. The latter case is the most interesting for us, since the sliding mode control principle does not allow the system to have a unique solution in the backward time. Any point of sliding manifold can be reached from the outside or along a trajectory in the manifold. For this general framework, the sliding manifold can be introduced by the same Definition 2.1 (or Definition 8.1 for discrete-time systems) using a shift operator U, a properly adapted notion of a manifold in X and, for example, finite-time extinction property [14–16].

For strongly continuous semigroups of linear bounded operators with $U(t) \in \mathcal{L}$, where $\mathcal{L}$ is a space of linear bounded operators, finite-time extinction can be defined as follows: [16, 17]: $\|U(t)\|_{\mathcal{L}} = 0$ for $t > t_0$, where $t_0 > 0$ is a sort of settling time. It is worth stressing that finite-time extinction in the infinite-dimensional framework may demonstrate even linear evolution equations [8, 16, 18]. Development of sliding mode control principle for abstract evolution system based on semigroup approach appears to be an interesting theoretical problem. Some attempts in this field were already done in [19–22].

11.5 On Stochastic SM

The publications on SMC have been devoted to the stability analysis of a class of nonlinear discontinuous feedback systems containing *bounded deterministic uncertainties*. Since stochastic systems contain *unbounded perturbations* typically having Gaussian distributions, the direct extension of SM ideas, successfully applied for deterministic systems, should be considerably changed or at least constructively reconsidered. After earlier 80s, some publications (see [23], [3, Chap. 14]), [24–28]) dealing with nonlinear Itô-type stochastic systems appeared in different scientific journals. They studied such new effects as actuator nonlinearities, time-delay terms, control input with sector nonlinearities and dead zones, Markovian jump parameters, polynomial system over linear observations, and so on. It is important to notice that majority of the cited references dealt with the so-called *multiplicative noise* where its power was proportional to the norm a sliding variable decreasing to zero when process approached the desired sliding surface.

To describe arising problems in stochastic SM systems, consider the simplest controllable stochastic model given in the Itô form in some filtered probability space

$$\left. \begin{aligned} dx_{1,t} &= x_{2,t}dt, \\ dx_{2,t} &= |u(x_t, t) + f(x_t, t)]dt + \sigma_1(x_t, t)dW_{1,t}, \\ dy_t &= Cdx_t + \sigma_2(x_t, t)dW_{2,t}, \end{aligned} \right\} \qquad (11.3)$$

where

$$x_t := \left(x_{1,t}^\mathsf{T}, x_{2,t}^\mathsf{T}\right)^\mathsf{T} \in R^{2n}, \ y_t \in R^m$$

and

$$\sigma_1(x_t, t) \in R^{n \times n}, \ \sigma_2(x_t, t) \in R^{m \times n}$$

are diffusion matrices of the standard n-dimensional Wiener processes, i.e.,

$$E\{W_{it} \mid \mathcal{F}_{t-0}\} \overset{a.s.}{=} 0, \ E\{W_{it} W_{it}^\mathsf{T} \mid \mathcal{F}_{t-0}\} \overset{a.s.}{=} t I_{n\times n}, \ i = 1, 2.$$

Function $f(x_t, t)$ is supposed to be bounded ($\|f(x, t)\| \leq L$) but unknown. Let the *sliding surface* $s(x) = 0$ is defined as

$$s(x) := x_2 + x_1 = \dot{x}_1 + x_1 = 0 \tag{11.4}$$

and the *sliding mode control* has the structure

$$\left.\begin{aligned}
u_t &= -Kx_t - k(x, t)\,\mathrm{sign}\,(s(x_t)), \ 0 \prec k(x_t, t) \in \mathbb{R}^1, \\
\mathrm{sign}\,(s) &:= (\mathrm{sign}s_1, \ldots, \mathrm{sign}s_n)^\mathsf{T}, \ 0 < K \in \mathbb{R}^{n\times 2n}.
\end{aligned}\right\} \tag{11.5}$$

The strong solutions of the closed-loop system are considered, see [29]. We intend to design K and $k(x, t)$ in (11.5) which guarantee the stabilization of the system (11.3) in a μ-neighborhood of the sliding surface $s(x) = 0$ (11.4) in some probabilistic sense.

The partial case when $m = n$, $C = I_{n\times n}$ (all states are observable) and $\sigma_2(x_t, t) = 0$ is considered in [30]. As shown in [30], selection

$$\left.\begin{aligned}
u_t &= -x_{2,t} - k(x_t, t)\,\mathrm{SIGN}\,(s(x_t)) \\
k(x_t, t) &= L\varphi_\varepsilon\left(\|x_{1,t} + x_{2,t}\|\right) + \frac{k_0}{2}\|x_{1,t} + x_{2,t}\| \\
\varphi_\varepsilon(\|s\|) &:= \begin{cases} 1 & \text{if } \|s\| > \varepsilon \\ \varepsilon^{-1}\|s\| & \text{if } \|s\| \leq c \end{cases}, \ \varepsilon > 0
\end{aligned}\right\} \tag{11.6}$$

guarantees the exponential μ-zone (with $\mu = \dfrac{\mathrm{tr}\{\sigma\sigma^\mathsf{T}\} + 2L\varepsilon}{2k_0}$) convergence for the second moment of the state, i.e.,

$$\left[E\{\|s(x_t)\|^2\} - \mu\right]_+ \leq Ce^{-k_0 t} \to 0 \text{ when } t \to \infty.$$

The *stochastic analog* of the deterministic *super-twisting control* may be expressed as

$$\begin{cases} dx_t = \left[-\alpha\sqrt{|x_t|}\,\mathrm{sign}x(t) + y_t\right]dt \\ dy_t = \left[f(x_t, t) - \beta\mathrm{sign}x_t\right]dt + \sigma dW(t) \end{cases}$$
$$|f(x_t, t)| \overset{a.s.}{\leq} L, \ \sigma > 0, \ \alpha = \alpha(x, y, t), \ \beta = \beta(x, y, t).$$

As it is shown in [31], if we select

$$\alpha = \alpha_0, \ \beta = \beta_0\varphi_\varepsilon(x) + \beta_{ad}$$

with α_0, β_0 such that

$$\min_{\alpha_0 > 0, \, \beta_0 > 0} \left\{ \frac{\alpha_0}{2}, \frac{\alpha_0 \beta_0}{1 + \alpha_0} \right\} > L, \ \alpha_0 (\beta_0 - L) > L, \ \alpha_0 > 4L$$

and

$$\beta_{ad} = -\frac{s_t}{|v_t| + \varepsilon_t} \text{sign} (v_t) \, \text{sign} x,$$

where

$$v_t = 2y_t - \alpha \sqrt{|x_t|} \text{sign} (x_t), \ s_t = k \sqrt{V_t} - \theta V_t,$$
$$V_t = 2\beta |x_t| + \tfrac{1}{2} y_t^2 + \tfrac{1}{2} \left[y_t - \alpha \sqrt{|x_t|} \text{sign} (x_t) \right]^2,$$
$$\varphi_\varepsilon (|x|) := \begin{cases} 1 & \text{if } |x| > \varepsilon > 0, \\ \varepsilon^{-1} |x| & \text{if } \ |x| \le \varepsilon, \end{cases} \ , \ \varepsilon_t = \begin{cases} \text{any} > 0 \, \text{if } s_t + \epsilon \ge 0, \\ \epsilon \dfrac{|v_t|}{|s_t + \epsilon|} & \text{if } s_t + \epsilon < 0, \end{cases}$$

we can guarantee the *mean-square exponential convergence* of V_t in the prespecified zone $\mu = \dfrac{\sigma^2 + \epsilon}{\theta}$, namely,

$$[\mathrm{E} \{V_t\} - \mu]_+^2 = O \left(e^{-2\theta t} \right) \xrightarrow[t \to \infty]{} 0, \ [z]_+ := \begin{cases} z & \text{if } z \ge 0, \\ 0 & \text{if } z < 0. \end{cases}$$

Possible applications of other Lyapunov's functions V_t are also discussed in [31].

As it was expected, sliding mode cannot exist in the traditional deterministic sense, governed by stochastic differential equations similar to (11.3). For the proposed control algorithm (11.6), we can speak about asymptotic convergence of the mean square value to vicinity of manifold $s = 0$. One more interesting aspect is the substantiation of validity of sliding mode equations using stochastic differential equations based on the regularization method (Sect. 2.6). It implies that due to small imperfections solution exists in the conventional sense and state trajectories are not confined to manifold $s = 0$, but run in a boundary layer, tending to zero if imperfections tend to zero. This limiting procedure results in equivalent control method equations for deterministic affine systems. Stochastic noise added to function s may be handled as small imperfection, if the gain multiplying noise tends to zero. At the intuitive level, this new regularization method should result in the equivalent control method equation as well. However, the condition "the trajectories run in a boundary layer" is not true for stochastic systems. Open problem: show the validity of the equivalent control method equations by this new regularization method.

Summarizing the results above, we can state that for the additive stochastic noise the special design (11.6) of the sliding mode gain parameters (in fact, linearly depending on the norm of the sliding variable s) may guarantee the exponential convergence of the averaged squared norm of the state vector to a μ-zone (μ-neighborhood) around the sliding surface $s(x) = 0$. This zone turns out to be exactly proportional to the diffusion parameter σ in the model description. Certainly, stochastic systems with SM controllers require very profound further investigation. As one can see, the system (11.3) with controller (11.5) should be considered as a stochastic differential inclusion. The existence of the solution for such system is also discussed in [30].

Several versions, such as

- the equivalent control method for stochastic SM,
- stochastic twisting and nested controllers,
- stochastic integral SM, stochastic SM observers,
- output-based SM with stochastic perturbations, and
- adaptive stochastic SM,

studied in [3, 32–34], require the extensions for the class of stochastic systems. In particular, the system with super-twisting control is discussed in ([31]).

References

1. Hsu, L., Kaszkurewicz, E., Bhaya, A.: Matrix-theoretic conditions for the realizability of sliding manifolds. Syst. Control Lett. (2000)
2. Filippov, A.F.: Differential Equations with Discontinuous Righthand Sides. Kluwer Academic Publishers (1988)
3. Utkin, V.: Sliding Modes in Control and Optimization. Springer, Berlin (1992)
4. Polyakov, A.: On settling time function and stability of vector relay systems. In: 12th IEEE International Workshop on Variable Structure Systems, pp. 149–154 (2012)
5. Zhuk, S., Polyakov, A., Nakonechnyi, O.: Note on minimax sliding mode control design for linear systems. IEEE Trans. Autom. Control **62**(7), 3395–3400 (2017)
6. Polyakov, A., Poznyak, A.: Invariant ellipsoid method for minimization of unmatched disturbance effects in sliding mode control. Automatica **47**(7), 1450–1454 (2011)
7. Osuna, T., Orlov, Y.: l_2-gain analysis of sliding mode controllers. In: Workshop on Variable Structure Systems. Nantes, France (2014)
8. Perrollaz, V., Rosier, L.: Finite-time stabilization of 2x2 hyperbolic systems on tree-shaped networks. SIAM J. Control Optim. **52**(1), 143–163 (2014)
9. Orlov, Y., Utkin, V.: Unit sliding mode control in infinite-dimensional systems. J. Appl. Math. Comput Sci **8**, 7–20 (1998)
10. Polyakov, A., Efimov, D., Fridman, E., Perruquetti, W.: On homogeneous distributed parameter systems. IEEE Trans. Autom. Control **61**(11), 3657–3662 (2016)
11. Orlov, Y., Pisano, A., Sodina, S., Usai, E.: On the lyapunov-based second order sm for some classes of distributed parameter systems. IMA J. Math. Control Inf. **29**(4), 437–457 (2012)
12. Pazy, A.: Semigroups of Linear Operators and Applications to Partial Differential Equations. Springer, Berlin (1983)
13. Krasnoselskii, M.A.: Shift Operator for Differential Equations. Nauka, Moscow (1966)
14. Sabinina, E.S.: A class of non-linear degenerating parabolic equations. Soviet Methematics Doklady **148**, 495–498 (1962)
15. Galaktionov, V.A., Vazquez, J.L.: Necessary and sufficient conditions for complete blow-up and extinction for one-dimensional quasilinear heat equations. Arch. Rat. Mech. Anal. **129**(3), 225–244 (1995)
16. Balakrishnan, A.V.: Superstability of systems. Appl. Math. Comput. **164**, 321–326 (2005)
17. Gohberg, I.C., Krein, M.G.: Theory and Applications of Volterra Operators in Hilber Space. American Mathematical Society, Providence, Rhode Island (1970)

18. Creutz, D., Mazo, M. Jr., Preda, C.: Superstability and finite time extinction for c_0-semigroups (2013). arXiv:0907.4812v4
19. Levaggi, L.: a). Infinite dimensional systems sliding motions. Eur. J. Control **8**, 508–518 (2002)
20. Levaggi, L.: b). Sliding modes in banach spaces. Differ. Integr. Equ. **15**, 167–189 (2002)
21. Orlov, Y., Dochain, D.: Discontinuous feedback stabilization of minimum-phase semilinear infinite-dimensional systems with application to chemical tubular reactor. IEEE Trans. Autom. Control **47**, 1293–1304 (2002)
22. Yan, X., Spurgeon, S., Edwards, C.: Variable-Structure Control of Complex Systems. Springer, Berlin (2017)
23. Drakunov, S.V.: On adaptive quasioptimal filter with discontinuous parameters. Autom. Remote Control **44**(9), 1167–1175 (1983)
24. Niu, Y., Ho, D.W.C.: Design of sliding mode control for nonlinear stochastic systems subject to actuator nonlinearity. Control. Theory Appl., IEE Proc. **153**(6), 737–744 (2006)
25. Niu, Y., Liu, Y., Jia, T.: Reliable control of stochastic systems via sliding mode technique. Optim. Control. Appl. Methods **34**(6), 712–727 (2013)
26. Wu, L., Ho, D.W.C.: Sliding mode control of singular stochastic hybrid systems. Automatica **46**(4), 779–783 (2010)
27. Gao, L., Wu, Y.: Control for stochastic systems with markovian switching and time-varying delay via sliding mode design. Math. Problems Eng. (2013)
28. Basin, M., Rodriguez-Ramirez, P.: Sliding mode controller design for stochastic polynomial systems with unmeasured states. IEEE Trans. Ind. Electron. **61**(1), 387–396 (2013)
29. Da Prato, G., Frankowska, H.: A stochastic Filippov theorem. Stoch. Anal. Appl. **12**(4), 409–426 (1994)
30. Poznyak, A.: Sliding mode control in stochastic continuos-time systems: μ-zone ms-convergence. IEEE Trans. Autom. Control **62**(3), 863–868 (2017)
31. Poznyak, A.: Stochastic super-twist sliding mode controller. IEEE Trans. Autom. Control **63**(5), 1538–1544 (2018)
32. Polyakov, A., Poznyak, A.: Reaching time estimation for super-twisting second order sliding mode controller via lyapunov function designing. IEEE Trans. Autom. Control **54**(8), 1951–1955 (2009)
33. Plestan, F., Shtessel, Y., Bregeault, V., Poznyak, A.: New methodologies for adaptive sliding mode control. Int. J. Control **83**(9), 1907–1919 (2010)
34. Utkin, V.I., Poznyak, A.S.: Adaptive sliding mode control with application to super-twist algorithm: Equivalent control method. Automatica **49**(1), 39–47 (2013)

Chapter 12
Conclusions

1. **Mathematical methods**
   All publications in this area since 50s until now can be separated into two groups:

   - Sliding mode equations are postulated.
   - Validity of equations is substantiated by regularization methods.

   The universal definitions of sliding manifold and sliding domain are discussed in the book; it is demonstrated that generally speaking sliding mode equations cannot be derived unambiguously and the cases when it is possible are singled out.

2. **Design procedures**
   The design methodology for SMC is the same in majority of publications:

   - A discontinuity manifold is selected such that the sliding motion of a reduced order exhibits the desired properties.
   - Discontinuous control is selected to enforce the sliding mode.
     Although dozens of publications can be found in the framework of this methodology, all of them are under the umbrella of the following modifications: component-wise discontinuous control, unit control, integral control, and high-order sliding mode control. The design ideas of each of them are explained in our paper. Selection of discontinuity manifold is a reduced-order conventional problem of the control theory. Enforcing sliding mode is equivalent to a stability problem also of a reduced order. The ideas of designing Lyapunov functions for each modification are explained.

3. **Lyapunov Stability Tools**
   Different forms of Lyapunov functions along with the homogeneity concept are offered to guarantee finite-time convergence, estimate the convergence time, and provide the desired convergence time by a proper choice of control parameters.

© The Author(s), under exclusive license to Springer Nature Switzerland AG 2020   125
V. Utkin et al., *Road Map for Sliding Mode Control Design*,
SpringerBriefs in Mathematics,
https://doi.org/10.1007/978-3-030-41709-3_12

4. **High-order sliding mode control**
   The new phenomenon was discovered: $(n - k)$-dimensional manifold $(k > 1)$ is reached after a finite time interval, and then k-dimensional sliding mode, called HOSM, occurs. HOSM is intended to overcome the problems related to relative degree and chattering.

5. **SM observers**
   All sliding mode observers in numerous publications are based on the same idea: enforcing the sliding mode along the observer input signal, specified as a discontinuous function of the observation error. As shown in our book, the majority of publications demonstrate that

   – order of the system is reduced;
   – the state vector can be found for time-varying systems;
   – the class of systems can be found such that the state vector is appropriately estimated in the presence of unknown disturbances.

6. **Discrete-time systems**
   Direct implementation of discontinuous control by discrete-time controller leads to chattering in some vicinity of the discontinuity manifold. The main attention was paid to analysis of this motion in majority publications in this area within last three decades. The recently introduced new concept "discrete-time sliding mode control" establishes the bridge between discrete- and continuous-time systems, since both of them have the same property: the state reaches a manifold in the state space and then the trajectories do not leave this manifold.

7. **SMC in infinite-dimensional systems**
   The chapter explains the reasons why the mathematical methods related to sliding mode equations for ODE should be newly developed. The new methods of deriving sliding mode equations and reaching conditions, based on Lyapunov's functional, for equations in Hilbert space are demonstrated. Practically, they embrace all particular cases.

8. **Chattering analysis**
   Chattering is the main obstacle for implementation of sliding mode control. It explains high intensity of research and publications on development of chattering suppression methods. Again each of them belongs to one of the following directions:

   – Systems with observers preserving a high-frequency component in control.
   – State-dependent amplitude of discontinuity in control.
   – Filtering out high frequencies based on the harmonic cancellation principle.

9. **Adaptive SMC**
   The objective of designing adaptive SMC is to reduce chattering following the second direction of the previous section. One of the two approaches can be found in each publication on adaptive SMC:

– Amplitude of control depends on the distance to discontinuity manifold.
– Amplitude of control depends on the equivalent control found easily by a low-pass filter.

The transparent explanations for both design ideas are given.

10. **Open Problems in SMC theory**

Interesting research directions are listed only, although not for all of them, even a solution idea can be outlined.

Printed in the United States
By Bookmasters